**SH**
**AN AMATEU**

## Other Titles of Interest

| | |
|---|---|
| BP89 | Communication (Elements of Electronics – Book 5) |
| BP91 | An Introduction to Radio DXing |
| BP105 | Aerial Projects |
| BP125 | 25 Simple Amateur Band Aerials |
| BP132 | 25 Simple Shortwave Broadcast Band Aerials |
| BP136 | 25 Simple Indoor and Window Aerials |
| BP145 | 25 Simple Tropical and MW Band Aerials |
| BP198 | An Introduction to Antenna Theory |
| BP255 | International Radio Stations Guide (Revised Edition) |
| BP257 | An Introduction to Amateur Radio |
| BP278 | Experimental Antenna Topics |
| BP281 | An Introduction to VHF/UHF For Radio Amateurs |
| BP290 | An Introduction to Amateur Communications Satellites |
| BP293 | An Introduction to Radio Wave Propagation |
| BP301 | Antennas for VHF and UHF |
| BP304 | Projects for Radio Amateurs and S.W.L.'s |

# SETTING UP
# AN AMATEUR RADIO STATION

by

I. D. POOLE
B.Sc.(Eng.), C.Eng., M.I.E.E., G3YWX

BERNARD BABANI (publishing) LTD
THE GRAMPIANS
SHEPHERDS BUSH ROAD
LONDON W6 7NF
ENGLAND

## Please Note



First Published – June 1991

British Library Cataloguing in Publication Data

Poole, I. D.

Setting up an amateur radio station.

1. Radio

I. Title

621.38416

ISBN 0 85934 245 X

Printed and bound in Great Britain by Cox & Wyman Ltd, Reading

# Preface

Amateur radio and shortwave listening in general is a fascinating hobby. However, setting up the station in the first place is not always as easy as it may seem. There are a whole host of decisions to be made. Everything from which equipment to buy and where to put the shack and make the best use of the available space to aerials which are very important, of course, because no station can operate properly with a poor one.

Many people like building some of their equipment. Sometimes a kit is a convenient route to use. At other times a circuit from a magazine may be built. In fact it may even be adapted to suit one's exact requirements. To do this some knowledge of the methods of construction which are open to the experimenter are necessary. In addition to this some test equipment may also be needed. Fortunately it is possible to cut down on the amount of test equipment required if the right techniques are used.

Having set up the station and used it for a while many people will want to be able to transmit. This opens a whole new dimension to the hobby as it is possible to contact people around the world or chat to friends on the other side of town. It also opens a completely new field for the constructor and experimenter. Today there are several different types of licence available. In the U.K. there are the standard class A and B licences. Then on top of this there are the new novice licences launched at the beginning of 1991. Decisions have to be made about how to set about studying for the tests needed to obtain the licence and which one to go for.

The aim of this book is to give some guidance in these decisions, and to help the newcomer to gain some experience from the mistakes of others without having to find out the hard way too much.

*I. D. Poole*

# Contents

Page

Page

# Chapter 1

# INTRODUCING AMATEUR RADIO

Amateur radio is a fascinating hobby which has been enjoyed by countless people since radio was first discovered. To many people it is a hobby purely to be enjoyed as a pastime. To many others it is a hobby which has developed into an interesting and rewarding career. In fact whatever the attraction of amateur radio, it is a hobby which is guaranteed to provide interest and relaxation.

**The Different Facets of Amateur Radio**

To some people the sheer enjoyment of listening is what attracts them to the hobby. On the bands it is possible to hear stations from anywhere in the world. Some stations may be quite close, in the same country or possibly even in the same town or locality, whereas other stations can be a considerable distance away where everything from the culture to the landscape and weather are totally different. This wide variety of backgrounds makes the long distance bands particularly fascinating.

Then in addition to this there is the variety in the types of stations which one can listen to. Some people enjoy listening to broadcast stations, hearing the programmes about their different countries, or news programmes from a different point of view. It can be quite illuminating listening to reports from two opposing countries.

Alternatively other people enjoy listening to the stations on the amateur bands. Here there is a wide variety of conversations which can be heard. Every manner of conversation from technical topics to quick exchanges of reports or discussions about the weather. In fact it is quite surprising how interesting some of the conversations can be.

Another dimension has been added to short wave listening with the introduction of scanners. These receivers can cover a very wide range of frequencies sometimes extending right up to 1 GHz or more. By using these receivers a whole variety of transmissions can be heard. Everything from aircraft and their traffic controllers to the modern cell phones. However when using these receivers it is necessary to be mindful of the law of the relevant country because it can be illegal to listen to some frequency bands or transmissions. Even so, where these transmissions can be heard quite legally they can be very interesting.

Apart from purely listening on the bands many people want to develop their hobby so that they can obtain an amateur transmitting licence. This adds extra interest to the hobby. It enables one to talk to people all over the world. There is the excitement of trying to contact amateur stations in far off countries. There is also the benefit of being able to talk to friends and people of similar interests whether they be on the other side of the world or the other side of town.

A transmitting licence also adds a new dimension to the experimental side of the hobby. It is possible to carry out all manner of experiments. Sometimes it may be trying out a new circuit or piece of equipment, other times it may be another form of experimentation. For example, many radio amateurs today take a great interest in the state of the ionosphere and make notes about the propagation. It is a well known fact radio amateurs have made a valuable contribution to our knowledge about propagation and many other areas of radio over the years.

Another aspect of the hobby which many people enjoy is construction. It can be a great challenge to build some of one's own equipment. Sometimes it may be a piece of ancillary equipment whereas other times it may be a transmitter. Whatever it is there is a great sense of achievement when it is complete and working. Many people have felt the excitement of holding their first contact on a new transmitter they have just built. Other people have been able to look at various gadgets around the shack and have been able to proudly say that they have built them.

There are many aspects to amateur radio. Not all of them will appeal to everyone. It may be that operating is the main interest whilst construction is undertaken only when necessary, or it may be the other way round. However, whatever one's interests there will always be a measure of practical skill required in setting up and operating the station.

**First Steps**

It may seem that it is quite difficult for the beginner to get started in amateur radio. After all the hobby has its own jargon associated with it and there can be quite a few technical aspects to the hobby as well. Fortunately these difficulties need not be a major problem. It is not necessary to be able to master all the technical aspects associated with radio. In fact there are very many radio amateurs who would not consider themselves in the least bit technical and yet they still manage to enjoy amateur radio to the full. However, in order to overcome the difficulties for the newcomer there are a number of steps which can be taken.

The first is to subscribe to one of the many amateur radio magazines which are available today. By doing this it is possible to keep up-to-date with what is happening within the hobby. In addition to this many of the magazines run features and series devoted to the beginner. It is also possible to gain a lot of information from the more advanced articles. Although they will not be fully understood at first small pieces will slowly start to fit into place, and it is surprising how quickly it is possible to pick up new ideas and concepts.

Apart from the magazines which are available there is a wide selection of good books. Some of them are available from High Street booksellers, but a much wider selection is available at most amateur radio retailers whose addresses can be found in the magazines. A good number of these books are aimed at the newcomer and can be very useful. Also there is a good selection of the more advanced books, and it is very useful to build up a reference library for day-to-day use.

Another suggestion is to join the national radio society. In Britain it is the Radio Society of Great Britain whose address is given in Appendix 1. The R.S.G.B. provides a wide range of services which are very useful to anyone starting out in the hobby as well as the more established enthusiasts. They publish a monthly magazine called *Radio Communication*. They also run a QSL bureau for people wanting to exchange QSL cards. Then they issue operating awards, run contests, give advice and negotiate for radio amateurs over licensing matters. In addition to this they publish and supply a very wide range of books.

Apart from joining the national radio society, it is also worth joining a local radio club. There are a large number dotted all over the country and usually there is one within easy travelling distance. Details of these clubs are often found in the radio magazines and the R.S.G.B. would be able to supply information about those in Britain.

There are very many advantages to be gained by joining a club. Most clubs meet quite regularly and have an active diary of lectures and events which can be both instructive and enjoyable. They also give the opportunity for the newcomer to talk to more experienced listeners or transmitting radio amateurs and take advantage of their experience. Usually club members will be most helpful to the beginner and can help them in a large number of ways.

**Decisions to be Made**

When starting out in amateur radio there are a large number of things to be planned. The most obvious is the station itself where decisions have to be made about the location of the shack as well as items like the receivers and transmitters which will be needed. In addition to this the aerials have to be given some thought.

Also when constructional projects are undertaken there are a number of different methods which can be used. Each has its own advantages and can be used to its best effect in different circumstances. Test equipment also has to be considered. Knowing what is needed and how to use the equipment is very important when undertaking any constructional projects, or routine maintenance around the shack. In fact some basic items are a virtual necessity in any amateur radio shack.

**Further Reading**

"An Introduction to Amateur Radio" (BP257) by I. D. Poole, G3YWX. ISBN 0 85934 202 6. Publisher: Bernard Babani (publishing) Ltd.

"An Introduction to VHF/UHF for Radio Amateurs" (BP281) by I. D. Poole, G3YWX. ISBN 0 85934 226 3. Publisher: Bernard Babani (publishing) Ltd.

# Chapter 2

# RECEIVERS AND TRANSMITTERS

Deciding upon the right equipment for a short wave listening or transmitting station is not easy. The choice of equipment which is available these days is quite staggering. There are several different manufacturers each offering their own range of equipment which are usually very comprehensive. In addition to all of the new equipment there is also a very wide variety of second-hand equipment. In fact the choice is vast to the extent that it can be quite daunting to the newcomer. However, matters can be simplified by making a few basic decisions about the requirements of the equipment before looking at anything.

**New or Second-hand**

One of the first considerations is whether to buy new or second-hand. It is obviously much nicer to buy something new, but the realities of life often mean that the cost of new equipment cannot be justified. In this case the second-hand option can solve the problem. There are a number of pitfalls of which any prospective buyer should be aware. First of all be careful in what is bought. Take time to sit down and try out the equipment properly. Anyone selling equipment who does not allow anyone to have a good look at the equipment may well have something to hide.

When trying the equipment out ensure that all the controls operate correctly, and all the switches work properly. If it is a receiver or a transmitter, check that it does not drift. Switch operation and drift often give a good indication about the overall condition of the equipment. If the switches do not operate very well then it is a sign that it has been well used and there could be other problems. Drift of the local oscillator is also important. It is not very common on the new synthesized rigs but it is fairly common on some of the older equipment. Obviously a certain amount is acceptable on very old items which only have a single conversion. However, on equipment which is not so old and has two or more frequency conversions with a crystal controlled first conversion should have very little drift indeed. Unfortunately it can be quite difficult to cure and very annoying during long periods of operation.

Also look for other tell-tale signs. Everything down to the appearance of the equipment. If it looks well cared for then the chances are that the owner has been careful. Even so, take into consideration the fact that it may well have been cleaned up ready to sell. However if this has been done then there are usually some visible signs. If it has been cleaned then it does not mean that it is not a good buy, but this has to be taken into account when assessing its real condition.

When buying second-hand equipment servicing is another very important factor. However reliable equipment is these days, there is always the possibility that something will go wrong at one time or another. When the equipment is new it can always be returned to the dealer, but for older equipment it is often a different matter. For example, there are still a lot of the pieces of equipment on the market which were manufactured fifteen or more years ago. Many of them are now no longer supported by their dealers and distributors, and often it can be a matter of do it yourself, particularly with the much older items. For valve equipment it should be remembered that even though most valves are still available, they are becoming increasingly more expensive.

Having described some of the drawbacks with second-hand equipment it is necessary to mention the advantages. The main one is obviously the cost. Usually used equipment sells for well below the price of new equipment, and if the readers' advertisements in the amateur radio press are studied then there are usually some good bargains to be found. Another advantage is that it is possible to buy some of the more traditional types of equipment second-hand. This may be an advantage in some cases because many people do not like a radio crammed with all the latest gadgets such as computer control, memories, keypads and the like.

**Choice of Receiver**

For any short wave listening station the receiver is the central item. It will also likely be the most expensive piece of equipment. This means that it is worth considering very carefully what facilities are likely to be needed. It is no use paying out extra cash for facilities which are not needed. On the other hand there is no point in buying a receiver which will not meet the requirements after only a short period of listening. This can obviously be a problem when buying a first receiver. In this case it is worth trying to see if it is possible to borrow or at least use a receiver for a while to assess what is needed.

One of the first points to consider is what frequencies will have to be covered. It is possible to buy a general coverage receiver. Generally they cover frequencies from below 1 MHz to 30 MHz or possibly a little above. Most new receivers today are general coverage receivers but some older receivers which can be bought second-hand were only designed to cover the limited bands. For example, a large number of amateur band only receivers were designed and manufactured up until 1980 or so. A receiver like this naturally limits the type of transmissions which can be heard. However it does have the advantage that the dial bandspread is very good, and if amateur operation only is foreseen then it should not be a problem.

Other basic aspects of the receiver's performance should also be considered. The receiver should have sufficient sensitivity. Most HF receivers today will be able to pick up signals which are exceedingly weak. For an HF receiver this will generally be quoted in terms of a signal of a certain number of microvolts at the aerial input giving an audio output having a certain signal-to-noise ratio. Today one would typically expect a signal of around 0.25 microvolts to give a 10 dB signal-to-noise ratio. But beware when comparing specifications as one receiver may be specified in a slightly different way to another. For example, one might give a 20 dB signal-to-noise ratio whilst the other might use 10 dB.

When looking at VHF and UHF receivers the sensitivity is generally quoted in terms of a noise figure. Here the front end contributes the majority of the noise which is heard. Accordingly a tremendous amount of work is put into reducing the noise which it generates. Therefore noise figures of 1 dB or so might be expected on some of the better receivers.

On HF receivers the front end noise is normally masked by the noise picked up by the aerial and so most HF receivers have an inferior noise performance when compared with their VHF and UHF counterparts. To illustrate this, good HF receivers will often have noise figures of 8 or 10 dB.

Selectivity is also important. Sufficient bandwidths should be incorporated to cater for all the different modes to be used. For SSB a bandwidth of about 2.6 kHz is standard and for CW it is useful to have a 500 Hz narrow band filter, but this may not be essential. If AM is to be used then a wideband filter must be used. Generally a bandwidth of 5 kHz or more may be needed.

Apart from the basic bandwidths of the filters it is also necessary to look at the skirt response, or the performance of the filters either side of the pass band. Often a filter can be designed to have an adequate bandwidth near the top of its response. However it may not attenuate the signals sufficiently when they are off channel. A filter in a receiver may be specified to have a certain bandwidth at the –6 dB point and then another bandwidth at a point further down its response. Normally this is specified at –60 dB. It is this figure which is all important in determining the off channel filter performance. Usually this performance is determined by the number of "poles" in the filter and this is sometimes quoted in the technical literature. The more poles, the better the filter.

In some instances it may be necessary to consider the modes of transmission that will be needed. Most HF receivers will have the capability of resolving AM, SSB and CW but only comparatively few are able to resolve narrow band FM properly. This facility may be quite important if the receiver was to be used for listening to Citizens Band or the amateur transmissions at the top end of Ten Metres. It would also be an important consideration if it was to be used with a converter to cover the VHF or UHF amateur bands.

Other functions like computer interfaces may be important. Today many stations will have a computer involved in many of the functions of running the station. Some receivers will include computer interfaces and if this is a requirement it is worth checking that the computer and receiver will be able to communicate.

**Transmitters and Transceivers**

Many similar factors need to be considered when buying a transmitter or transceiver. Frequency coverage, cost and whether to buy new or second-hand will all need to be considered. Naturally there will be a number of other factors which will have to be considered as well.

The first, and probably most obvious, is the actual transmitter power. It is nice to be able to run plenty of power, but high power transmitters often cost more than the lower power ones. They are also more likely to cause interference to neighbouring radios, hi-fi systems, and televisions so this is an important consideration.

Whilst looking at the output capability of a transmitter it is worth looking at the duty cycle which it can handle. This is all important if some of the data modes which are gaining in popularity are to be considered as they require the transmitter to be able to operate with a high duty cycle. This needs to be taken into account because some of the older transmitters will not be able to run their full power for long.

The modes of operation is another consideration. Many people will only like to use CW on the HF bands, whilst others will want FM for VHF local communications. Most people will probably want to have the possibility of using all the commonly used modes for their favourite bands. This is less of a problem below 30 MHz where SSB and CW (together with various data modes) are the two major modes which are used. Above 30 MHz FM is also widely used. To meet this need multi-mode transmitters are available but obviously they are more expensive.

It is worth looking at the general facilities which the transmitter offers. For example, many units today include speech processors and CW keying circuits as standard. Another useful facility is to be able to operate on split frequencies. This can be important if DX chasing on the HF bands is to be considered. Here it is quite common practice for DXpeditions to operate on split channels to enable better operation through a large pile up. Sometimes the transmit and receiver frequencies may be 10 kHz or more apart.

**Converters and Transverters**

In order to extend the coverage of a receiver it is possible to add a converter. This can save the cost of having to buy a completely new receiver to cover a new band. One example of this is when a converter is used to convert signals on the two metre band down to a frequency which can be covered by a general coverage or amateur band HF receiver.

Generally these converters have a fixed frequency local oscillator so that the frequency is shifted by a fixed amount. For example a typical two metre converter will have a local oscillator frequency of 116 MHz. Therefore signals at 144 MHz will appear at 28 MHz on an HF receiver and signals at 146 MHz will appear at 30 MHz.

It is also possible to use the same type of idea for a transceiver where both transmitted and received signals are converted in frequency onto a new band. This type of unit is called a transverter. Again they are often used to extend the use of an HF transceiver onto VHF or UHF. However, there are also many transverters which are used to enable VHF or UHF transceivers to operate on higher frequency bands. Also with the large number of VHF transceivers around nowadays there is a growing number of transverters available which enable these transceivers to be used on the HF bands.

Whilst converters and transverters may not be as convenient as using a purpose-built unit for the extra band, they are much cheaper and enable the maximum use to be made of the equipment in the shack.

**Ancillary Equipment**

There are a number of small accessories which will almost certainly be needed after a while. Items like an aerial tuning unit to match up the aerial to obtain the best performance from it. Then for those with transmitting licences, a standing wave ratio (SWR) meter will always be useful to have around as it will give an indication of how the aerial is performing and warn of the possibility of a mismatch which might affect the transmitter.

Whilst these items are all fairly cheap to buy, they are also ideal items to build. Fortunately they are usually quite simple to put together and will always be useful. As such, they make ideal starter projects.

# Chapter 3

# SETTING UP THE SHACK

Apart from buying the equipment it is also necessary to have somewhere to keep it and use it. Traditionally the radio room has been called a shack, and often the term seems to describe the looks of it very well. However it must be said that some shacks are very tidy and well set out. Whatever one's version of a shack it is as well to put some thought into setting it up as this will pay dividends later by making it much more pleasant and easy to use.

There are many different places which can be used for a radio shack. Some people have been fortunate enough to be able to set aside a complete room and do it up accordingly. Many have had to make do with odd corners of the house which the rest of the family do not want. Even so these corners have been used to the full because of the thought and ideas which have been put into their planning. In fact time and effort put into planning the shack from the very beginning will pay dividends later as it will be possible to gain the most from it in terms of enjoyment, comfort and convenience.

**Requirements of the Shack**

Ideally the shack should be set aside from the rest of the living room in the house. This will mean that all the noise and paraphernalia that goes with the hobby can be contained and is less likely to spread all over the house. It will also mean that small children can be kept away from the equipment and there will be less danger to the children and also to the equipment itself!

Another requirement for the shack is that it should be large enough to contain everything. It is amazing just how much can be accumulated even over a short period of time. Space is needed for the equipment itself as well as books and all sorts of other items associated with the hobby. If it is at all possible all of this should be contained within the shack so that when they are needed they will be close at hand and easy to get hold of.

A mains supply is another obvious requirement. Most areas within a house will have mains sockets at hand. However if the shack is located somewhere else then a mains supply may not be available and one will have to be installed.

It should also be possible to route aerial feeders into the shack without too much difficulty. In most shacks this is not too much of a problem, but in certain locations it may mean routing unsightly feeders around the house. This is unlikely to meet with much approval from the rest of the household.

In some shacks it might be necessary to have a good earth connection. Aerials like longwires need an efficient earth system and the earth lead should not be too long.

**Location**

There are a number of different places which can be used as the basis for a good shack. The ideal is to set aside a separate room in the house. If this is not possible other alternatives have to be sought. Fortunately there are many different places which can be converted into good shacks and many people have made excellent shacks in some of the most unlikely positions. Often these places will have some disadvantages and these should be considered carefully before ploughing a lot of time, effort and money into any conversion project.

Obviously the ideal location for many people for their shack is a spare room in the house. An unused bedroom for example. It is very convenient because access is easy and it is possible to drop into it for five minutes or so to see what activity is like on the bands. In addition to this it can be heated easily and there will be mains outlets already installed. Access for feeders is normally not too difficult. Having at least one outside wall and windows, feeders can be run through the ceiling if the room does not have any further storeys on top of it.

Another possibility is to use the loft. Many people have successfully converted a loft space into a good shack as there is usually a lot of space there. However there are a few drawbacks to be borne in mind. The first is the cost of conversion. Depending upon the sort of job which is done, this might be comparatively little, or it could be very expensive if the joists have to be reinforced and a full loft conversion undertaken. If this is not done then the loft will undergo large temperature changes becoming very cold in the depths of winter and exceedingly hot at the height of summer.

Another possibility which a number of people have tried is to use a garden shed. This can be made into a very good shack. However it can mean a fair amount of work if it is to be done properly. The shed would have to be lined to give a degree of temperature insulation. Then items like the mains supply would have to be laid on. However, access to the aerials is normally very good.

Other people have used sections of garages. Again this can be made to give quite a reasonable shack if enough work is put into it. It can provide a well built structure to house the shack, but it is necessary to ensure that there are not too many draughts.

Some shacks have even been set up in a cupboard of one shape or another. In fact a cupboard can be made to give quite a good shack especially if it is well planned. Ideally it should be one of the larger walk-in variety which are built in to the house. They can be large enough to accommodate a table, the equipment, as well as a selection of components, books and the like.

**Table for the Shack**

One of the most important pieces of furniture in any shack will be the radio table. It can either be bought or made. In fact this decision will often be determined by the shape and size of the room and whether it is actually possible to buy a suitable table.

Whatever the decision about constructing or buying the table there are a few requirements which will need to be met. The first is that the table must be strong. It is surprising how quickly the weight of equipment builds up. This is even true for modern equipment which is quite light in comparison to the older units. When the weight of the transceiver or receiver, power supply, antenna tuning unit and all the other units are added up then it can place quite a strain on the table. If some older equipment is to be used then it is even more important to ensure that the table can withstand the load. This means that ordinary chipboard should not be used because it will sag under comparatively small loads. It even has to be very well supported if it is used for a shelf. Probably the ideal is to use ¾-inch blockboard or something similar as this will be strong enough for most equipment but even this may need a firm frame underneath it.

Another important requirement is to ensure that the table has sufficient depth. There should be enough space to allow for the equipment itself with two or three inches at the back and about eighteen inches in front. By allowing this much space in front it will be possible to accommodate a log book, notebook, microphone, morse key and so forth quite comfortably. This space will also be sufficient for constructing projects as well.

If the table top is to be attached to the wall, then enough space must be left behind it to get wires up and down. In fact it is best to leave enough space for the mains connectors as well because it is very annoying to have to take the connectors off to pass a lead up or down.

Another consideration is the height for the work surface. This should be high enough to enable any work being done on the table to be undertaken comfortably. Generally desks and the like are about two foot six inches from the ground but the optimum height will be dependent upon the chair in use. This means that it is worth trying to do some experimentation to find the best position. By doing this it could save backache later and make working in the shack more comfortable.

It is well worth considering finishing the work top off with Formica or something similar. This can be easily cleaned and is more likely to be resistant to damage than a plain wood finish.

**Shack Wiring**

It is worth planning the shack wiring from an early stage. It can be quite annoying if mains cables are strewn over the work surface. An ideal solution to wiring the mains is provided by the mains adapter blocks which accept about four plugs. One or two of these can be fixed to the back of the table and all the equipment leads passed behind the table. This keeps all the unnecessary wiring out of the way. It also means that a single lead can be taken from the table to one of the normal house mains outlets. One advantage of this is that a single switch can be used to isolate the whole of the station, and this reduces the possibility of accidentally leaving just one piece of equipment on.

A further advantage of having a single cable supplying the radio table is that a protective circuit breaker can be used. Today Residual Current Circuit Breakers (or RCCBs for short) are generally used as they can be installed quite easily and they give a good measure of protection. They come in various forms. Some are actually included in as part of the mains plug, whilst others are available to be fitted as an in-line item in the cable. RCCBs are quite easy to obtain these days and they can be obtained from most electrical shops.

Essentially an RCCB monitors the current in both live and neutral lines and if there is an imbalance of more than 30 milliamps or so it will trip out. This imbalance will occur if there is any current being drawn to earth which can happen if there is a short to earth or if someone touches a live wire.

It is obviously advisable to fit an RCCB. However, even when one is fitted the same care should be exercised when dealing with mains voltages. An RCCB will not prevent anyone from receiving a shock. It will only help prevent the effects from being too serious.

**Earthing Arrangements**

When dealing with any radio station the earthing arrangements are of great importance. They are of particular interest when dealing with a transmitting station as poor earthing arrangements and long earth interconnection leads between different units can lead to problems with RF instability and so forth. One way of overcoming this is to install a large earth braid or copper strip at the back of the table. Then all the different units can be connected to the earth strip by a short lead. By doing this all the earths of all the units will be properly connected together.

**Lighting**

The lighting within the shack is important. The radio table needs to be well illuminated so that constructional projects can be undertaken. Good lighting is also needed so that the equipment can be operated satisfactorily. For this the lighting should be above the table and not behind the operator. If this is done then he will not work in his shadow.

Lamps which can be angled and moved are ideal to provide extra light. They come in very useful for constructional projects because they can provide a high level of light where it is needed.

When choosing the main light for the whole of the shack careful consideration should be given to the sort which is to be used. It is found that fluorescent lights emit significant levels of RF up to very high frequencies. In many cases this might not be a problem, but it is worth avoiding them for the shack if at all possible.

**Equipment Layout**

Having built the shack it is worth giving thought to the actual layout of the equipment as this can contribute a lot to ease with which it can be used and hence the enjoyment of the whole shack. Although there are no hard and fast rules there are a number of guidelines which should be borne in mind when planning the layout and one similar to that shown in Figure 3.1 would be quite satisfactory.

The main receiver or transceiver should be placed centrally on the table with the tuning dial only a couple of inches above the table surface. This means that one's arm can rest on the table whilst the receiver is tuned. This saves a lot of arm ache during extended periods of operation.

A second receiver or transceiver can be placed to the right of the main one. With it placed in this position it is again easily accessible without having to reach across the table from right to left.

If a separate transmitter is used this should be placed to the left of the main receiver in the centre. This position is ideal because it enables the microphone to be held in the left hand whilst using the other hand for writing in the log book or making notes. If a separate transmitter is not used then this position can be used for a linear amplifier.

If a morse operation is envisaged then the key should be placed on the right hand side of the table. It is also worth

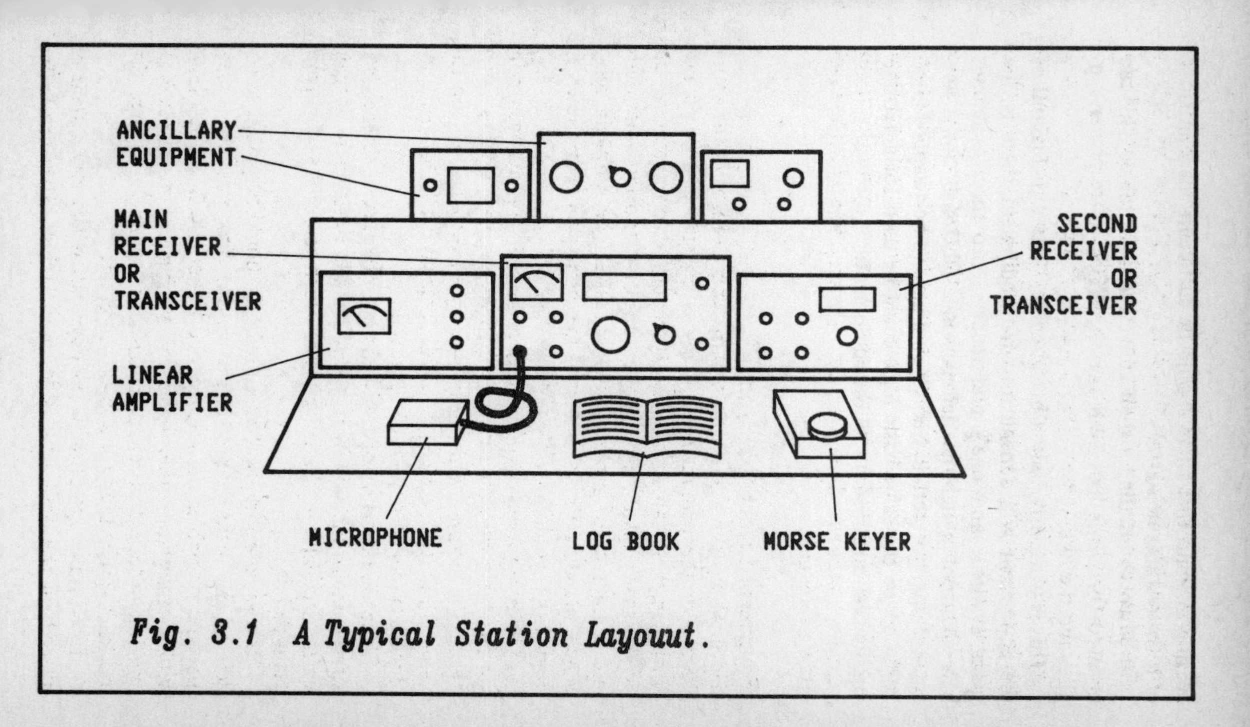


*Fig. 3.1 A Typical Station Layouut.*

bearing in mind that there should be sufficient space in front of it for resting one's arm.

It is worth building a shelf above the main table. This can be used for items like SWR meters, ATUs, spare power supplies and the like.

This is obviously only a broad guide. The actual layout of the equipment will depend on what equipment there is, the space available and one's preference. Also these guidelines have assumed a right-handed person. Obviously for a left-handed operator certain items would have to be swapped over, but even so the same basic logic could be used for determining the best positions for all the equipment.

# Chapter 4

# AERIALS

The performance of the aerial is crucial to the overall operation of the station whether it is used for receiving or transmitting as well. A good aerial will enable weaker and more distant stations to be heard or contacted. However, if the aerial is poor then it can sometimes be difficult to make many contacts. Any time and money invested in the aerial will always pay dividends. It is likely to be far more worth while than spending more money on updating the equipment in the shack to the very latest model.

As the aerial is so important the choice of the type of aerial and its location needs to be thought out carefully. Obviously most people would like a large aerial on a tall tower, but this is not usually possible. It is more likely that the aerial will have to be erected in a fairly small garden without having too much of a visual impact on the surrounding area. This restricts what can be done to a very large degree, and it calls for a great deal of ingenuity if the best compromise is to be found. In addition to this most people will not want to be repeatedly fitting up and trying out new aerials, so that whatever is put up will have to stay up for some while.

For any location or situation there will be a variety of aerials which can be used. Each type has its own merits, and its own disadvantages. For example one type of aerial may be better for short haul contacts whilst other types may be better for long distance ones. Again some may fit into small gardens better than others. Unfortunately it is not always easy to know in advance what is going to fit the bill best of all. However, with a little knowledge about the various types of aerial which are available it is quite possible to make a very good judgement of what will work well.

## Location

Before discussing the relative merits and disadvantages of the various types of aerial it is necessary to consider where the aerial can be erected as this will have a bearing on the choice of aerial.

Most people will be able to fit an aerial up outside. However there will always be some instances where this is not possible. Whilst using an internal aerial will not be nearly as good as having one outside, if this cannot be done then all is not lost. It is quite possible to hear and contact some very long distance stations although it will obviously be more difficult. When deciding on where to put the aerial, it should be placed as far away from any metalwork as possible. Usually an internal aerial will be in the loft and items like chimney liners, water tanks, as well as any electrical wiring should be avoided. This is particularly important where directional VHF and UHF aerials are concerned.

The other item to be aware of when using internal aerials is the possible health hazard if it is to be used for transmitting. It is now recognized that high RF fields can represent a health hazard. Accordingly aerials should be erected so that they do not direct power through inhabited parts of the house.

If the aerial is outside it should be as high as possible. This will help it to clear the screening effects of any nearby objects. It is surprising how even an ordinary house can act as a screen. The major reason for it acting as a screen is all the wiring and piping which there is. However, as the frequency rises even the brickwork and general structure of the house will attenuate the signals.

Another factor which is worth considering is the distance which the aerial is located away from the house. Within the average house a large amount of interference and noise is generated. Fluorescent lights, televisions, computers and many other domestic electrical and electronic appliances can generate a surprising amount of interference. If the aerial is moved away from the house then the effect of this will be reduced.

Moving the aerial away from any houses can have advantages when a transmitter is used as well. First of all it will ensure that there are no areas within the house where there are high levels of RF and secondly it will reduce the levels of any interference which might be caused by the transmitted signal to television and hi-fi systems. This can be particularly important when neighbours are suffering from interference.

## Aerials for H.F.

Although basic aerial theory holds true for any frequency the type of aerial which is used may depend upon the frequency in use. For example some aerials which are common at VHF and UHF frequencies will be far too large for use on the HF bands. Conversely some aerials which are used at HF are not practicable for use on the higher frequencies.

For most people the choice of an aerial for the HF bands will be a choice between some of the simpler styles of aerial like a "longwire", a dipole, or a vertical. Obviously some will have the space for a large beam, but by and large this type of aerial is the exception rather than the rule at these frequencies.

## The Longwire

The longwire is one of the most commonly used aerials in use, particularly by short wave listeners. Strictly speaking it should be called an end fed wire because a proper longwire is several wavelengths long at its frequency of operation. It then becomes quite directive with the main beams tending to align with the axis of the aerial. However most of the wires in use are comparatively short and are often less than a wavelength.

The end fed wire like the one shown in Figure 4.1 has gained popularity because it is very easy and convenient to use. Its length is not critical because it can be tuned to resonance with an aerial tuning unit (ATU) which will enable it to be used on a wide range of bands.

The tuning unit does not have to be very complicated. Either it can be bought as a ready-made item or alternatively it can become a constructional project. A whole host of designs exist in the amateur radio magazines.

In order to make the best of an end fed wire it should be a minimum of a quarter of a wavelength long. If it is any shorter then its efficiency will be reduced quite considerably even though it can be tuned to resonance and will give a low VSWR.

Another requirement for an end fed wire is that it should have a good earth system for it to work against. Ideally the earth lead should be as short as possible and the earth should have a low resistance.

Although end fed wires offer the ideal solution to many aerial needs they do have some disadvantages. One is that as

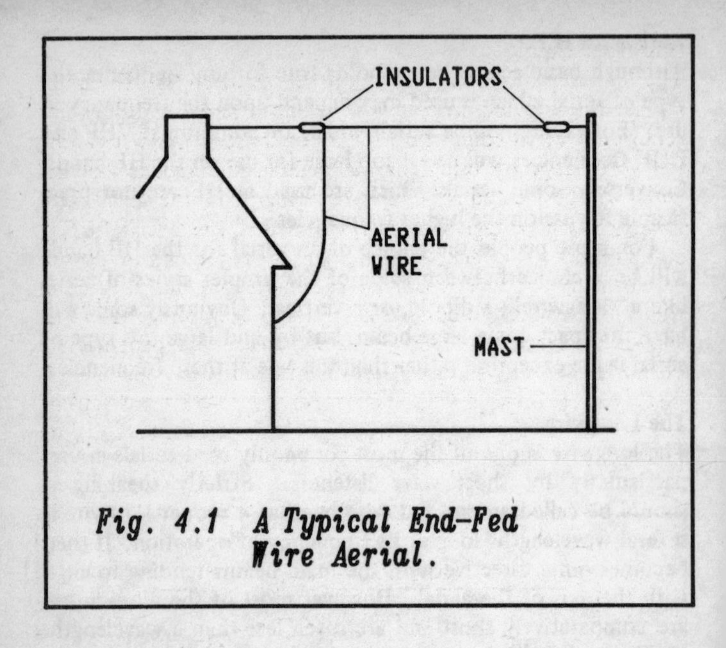


*Fig. 4.1 A Typical End-Fed Wire Aerial.*

soon as the wire leaves the aerial tuning unit then it will pick up and radiate signals. From the point of view of reception this means that any interference generated in or around the shack from items like lighting or computers will be picked up and impair the reception. From the transmitting standpoint energy will be radiated within the shack unless the tuning unit is located remotely. This will mean that there will be high levels of RF in the shack. In turn this can lead to a health hazard if high levels of RF are used. In addition to this, it can give rise to feedback caused by RF getting into the equipment. This can cause particular problems when modes like SSB are used.

Finally the earth for the aerial can often create some difficulties. If the shack is located above ground level then the earth lead will have to be fairly long. Although the aerial will still operate it will mean that the equipment will be above earth potential for RF signals. This will make the problems of RF in the shack still worse.

**Half Wave Dipole**

The half wave dipole is very frequently used on the HF bands. Unlike the end fed wire it is a balanced aerial and does not require an earth connection for its operation. Also it can be fed very easily by using coaxial feeder which means that the aerial can be located some distance away from the shack. This is a distinct advantage because it will reduce the levels of RF in and around the shack, so reducing the problems it causes.

Normally a dipole is half a wavelength long and it is fed in the centre where the current is high and the voltage is low as shown in Figure 4.2. This means that the feed impedance will be fairly low and in fact it works out that the impedance is just over 70 ohms in free space. However, this value will be altered to a degree by the height of the aerial and the proximity of other objects.

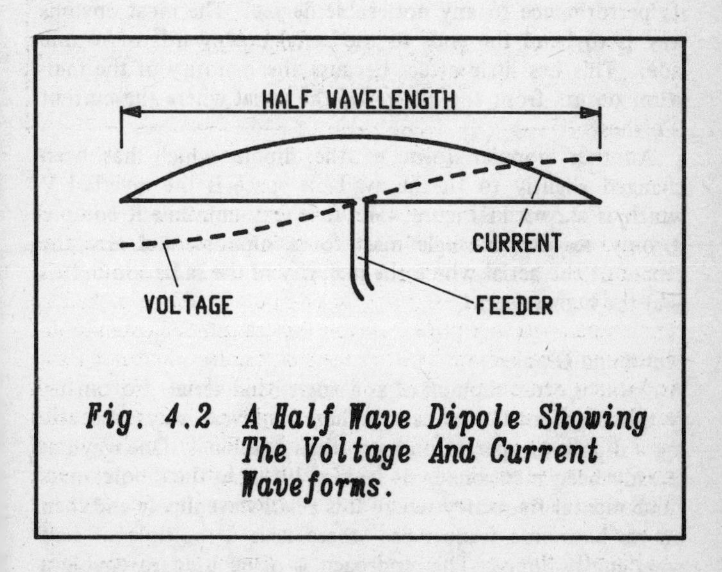


*Fig. 4.2 A Half Wave Dipole Showing The Voltage And Current Waveforms.*

Whilst dipole aerials are generally a half wavelength long there is nothing to stop them being made an odd multiple of half wavelengths in length. If this is done then it will present the same feed impedance as it is fed at a point of high current and low voltage. The major difference is that the polar diagram of the aerial will change. It is found that the direction of major radiation from a half wave dipole is at right angles to the axis of the aerial as shown in Figure 4.3(a). However, when the length of the aerial is increased there will be a number of different maxima or lobes. These tend to align themselves more and more with the axis of the aerial as the length is increased. In addition to this a number of minor lobes appear. Figure 4.3(b) shows the example of the three half wavelength dipole which is probably the most common example of an extended dipole.

Having suggested that an aerial can be made longer than the minimum length possible most people will be in the position that they want to decrease the size of an aerial to fit it into their garden. Fortunately it is quite possible to bend a dipole to make it fit the available space without degrading its performance to any noticeable degree. The most obvious way is to bend the ends of the aerial downwards or to one side. This has little effect because the majority of the radiation occurs from the centre of the aerial where the current is highest.

Another popular form of the dipole which has been changed slightly to fit the available space is the inverted V which is shown in Figure 4.4. It is very convenient because it only requires a single mast for a support, and also the centre of the aerial where the majority of the radiation occurs is at the highest point.

**Multiband Dipoles**

A dipole is often thought of as a single band aerial. Fortunately this need not be the case as there are several ways of enabling a dipole to operate on more than one band. One way has already been mentioned. It is done by using the dipole on its fundamental frequency where it is a half wave dipole and then on its harmonic frequencies where it is a multiple of half wavelengths long. This approach is often used to enable a forty metre dipole to operate on fifteen metres as well.

Another common approach is to connect several dipoles to a single feeder as shown in Figure 4.5. In this way a dipole which appears to the receiver or transmitter as a multiband dipole can be made quite easily. Although it may seem that

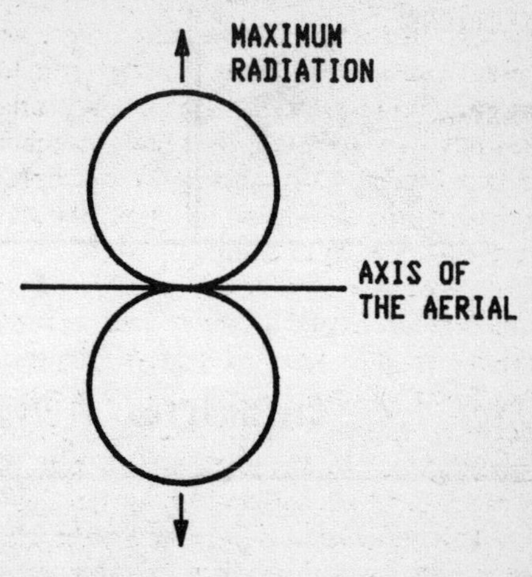


a] HALF WAVE DIPOLE.

b] THREE HALF WAVELENGTH DIPOLE.

*Fig. 4.3 Polar Diagrams For Dipoles Of Different Lengths.*

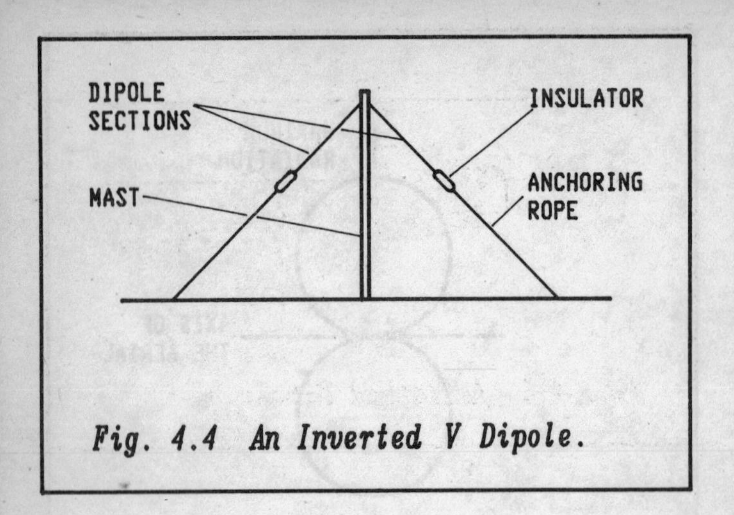


*Fig. 4.4 An Inverted V Dipole.*

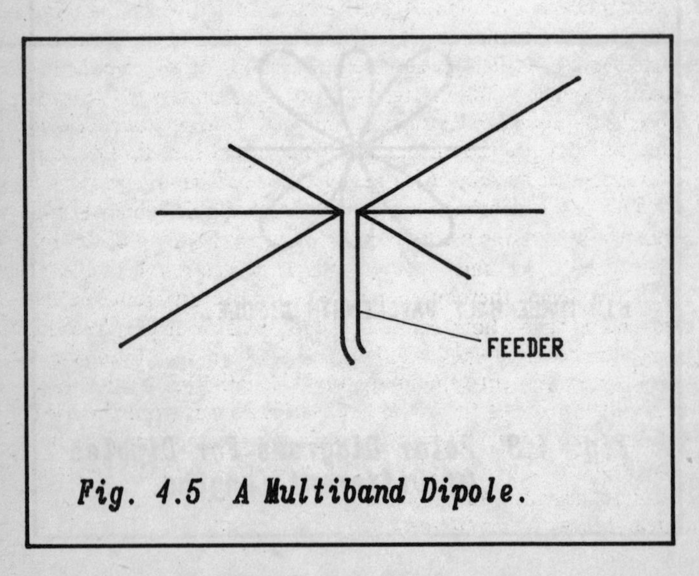


*Fig. 4.5 A Multiband Dipole.*

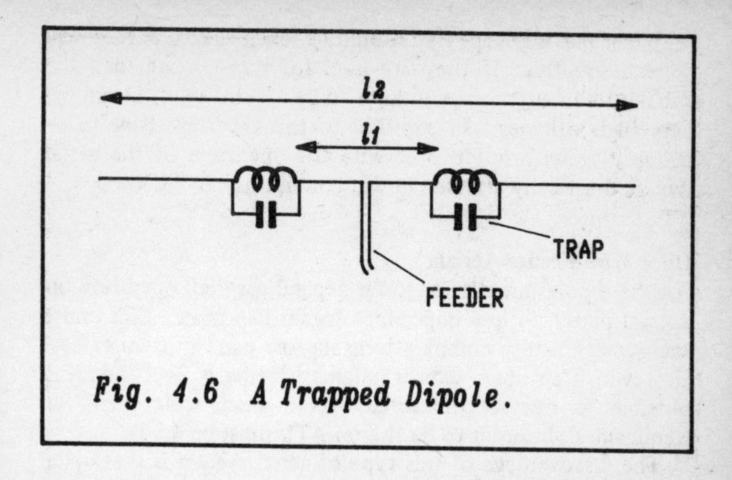


*Fig. 4.6 A Trapped Dipole.*

the different elements will impair the performance of the ones in use this is not so to any great extent. The reason is that they are not at resonance and therefore they present a high impedance to the feeder. Naturally there is a small effect between the elements and this is noticed when trimming the aerial to resonance. It is found that trimming a dipole for one band can slightly alter the resonance of the dipole for another band. This means that any trimming of the lengths should be done a bit at a time and with a complete check of the performance on all bands. This should save finding that too much has been cut off the length for one band.

The other approach to a multiband dipole which is frequently seen is the use of traps as shown in Figure 4.6. In this arrangement the traps are used to isolate a section of the aerial at a frequency thereby giving two or more different lengths depending upon how many traps are used. For example the traps in the aerial shown in Figure 4.6 would need to resonate at a frequency corresponding to the length $l_1$. In this way only the length of the aerial between the traps would be used at this frequency. Then at a frequency corresponding to the length $l_2$ the whole aerial would be operative. In fact the traps tend to have a loading effect and this will make the aerial a little shorter at this frequency.

When making traps care should be taken to ensure they will work correctly. If they are used for transmitting then the capacitors in the traps will have to be capable of withstanding very high voltages. In addition to this the traps have to be suitably water-proofed otherwise the operation of the aerial will be drastically affected in wet conditions.

**Open Wire Feeder Aerials**

All the dipoles mentioned so far depend for their operation on a good match to low impedance feeder like coax. This limits the operation to a comparatively narrow band of frequencies. However, if an open wire or balanced feeder is used then it is possible to operate the aerial over a much wider band of frequencies. In order to do this an ATU must be used.

The disadvantage of this type of aerial system is that open wire feeder is not as tolerant as coax to being run through the house or close to any objects which might unbalance it. However one advantage is that it is quite easy to make the feeder. This can be done by taking two lengths of wire and spacing them about 6 to 8 inches apart. To keep the feeder spacing constant, spacers must be placed along its length at intervals of about eighteen inches or so. Although it is possible to buy spacers they can be made quite easily in a number of ways. One idea is to collect a number of the plastic cases from ball-point pens. A hole can then be drilled in either end large enough to pass the wire through. Once on the wire a small kink is then sufficient to hold them in place.

As an alternative to the true open wire feeder it is possible to buy 300 ohm flat ribbon feeder. If this is going to be used then the clear plastic variety is not at all suitable for external use as it absorbs moisture and becomes very lossy in a damp or wet environment. The best cable is the black variety with holes in the dielectric as shown in Figure 4.7.

**Verticals**

Vertical aerials form another type of aerial which are very popular on the HF bands. Generally they are a quarter of a wavelength long because although longer aerials can be made the size generally limits the length.

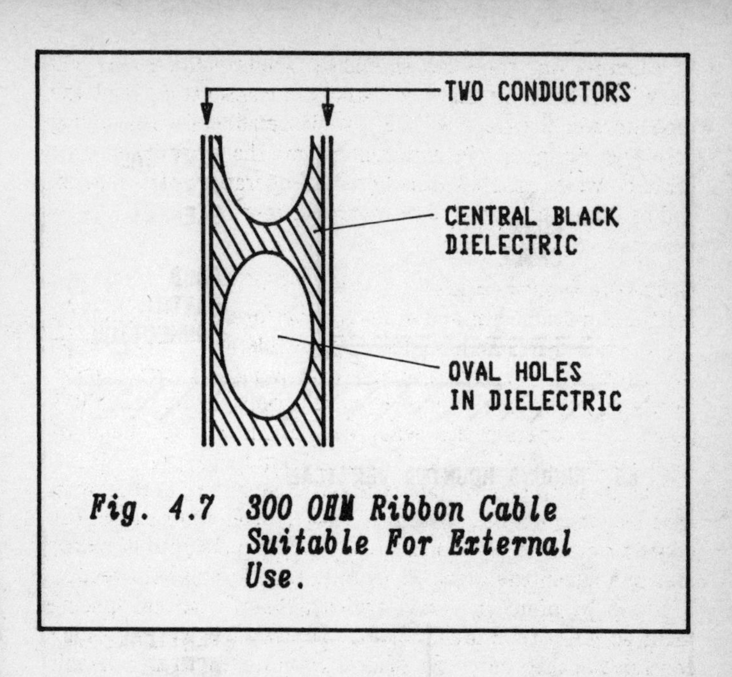


*Fig. 4.7 300 OHM Ribbon Cable Suitable For External Use.*

Whilst it is quite possible to make verticals one's self, there is a very wide range of commercially made trapped verticals which can be bought for around £100 or so. These aerials operate on the same basic principle as the trapped dipoles using tuned circuits to isolate sections of the aerial so that the correct electrical length is seen for the band in use. Often these trapped verticals will have two or more traps which enable them to cover a number of different bands.

Verticals are ideal in certain instances because they can be erected so that they take up very little space. If they are ground mounted, i.e., mounted at ground level using an earth connection instead of a set of radials as shown in Figure 4.8(a), they can be fitted into a very small garden.

When this method is used it is essential that the earth which is used is very good, offering less than an ohm or two. If the earth is not very good then the efficiency of the aerial will be markedly impaired. With this in mind it is worth checking that a good earth system can be installed before embarking

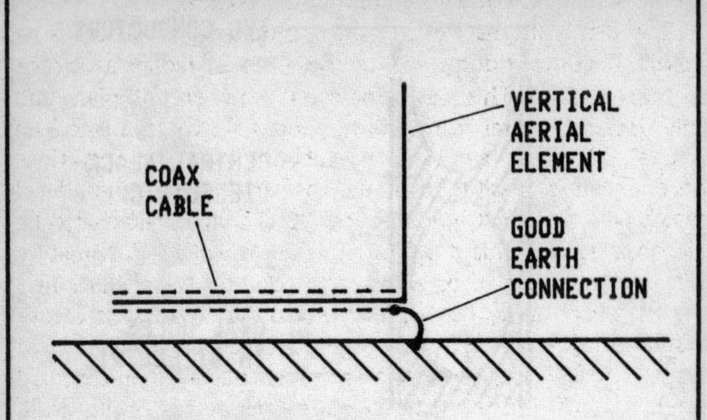


a] GROUND MOUNTED VERTICAL.

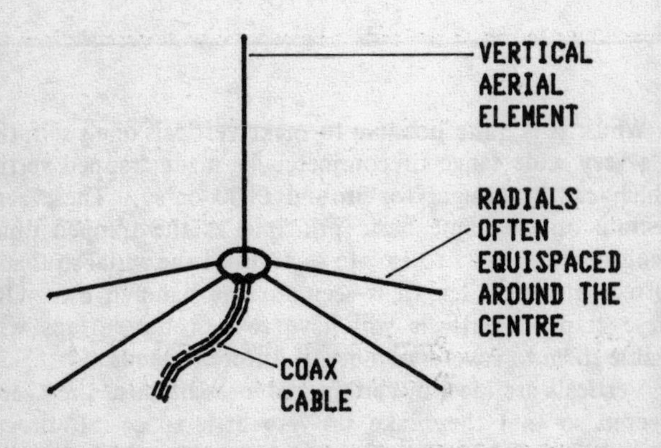


b] GROUND PLANE VERTICAL.

***Fig. 4.8 Ground Mounted And Ground Plane Verticals.***

on the installation of a ground mounted vertical.

The other alternative to ground mounting a vertical is to mount it above ground and to use a set of radials as shown in Figure 4.8(b). This makes the aerial a true ground plane and this option is better in several respects. Firstly, the aerial no longer relies on an earth connection for its operation. Secondly, it is possible to mount the aerial above ground level which will mean that it will not be screened by nearby objects. However against this a set of radials is needed. Fortunately this does not have to be a full set of quarter wavelength ones for each band in use. It is possible to obtain quite satisfactory results using a reduced set which can even be loaded or use traps. In fact some verticals are even sold with an optional single trapped radial which seems to operate reasonably well. This approach could be very attractive where space is at a premium.

In general verticals perform well as DX aerials. As shown in Figure 4.9 they have a low angle of radiation. This is an

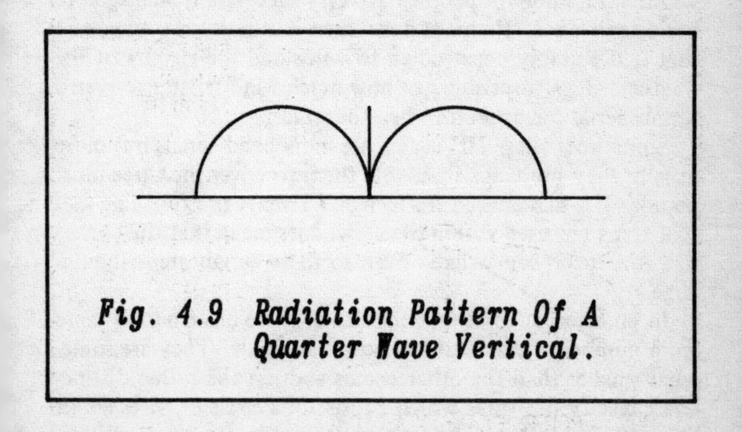

*Fig. 4.9 Radiation Pattern Of A Quarter Wave Vertical.*

advantage because not only does it bring in the long distance signals which tend to come in at a low angle but it cuts out much of the more local traffic which comes in at a higher angle. Unfortunately verticals are usually more expensive than some other aerials. A simple dipole can be strung up for next to nothing, but even if a vertical is home-built the materials needed to make it will be more expensive than the wire for a dipole. Even so the cost of the vertical may well be worth the added expense.

### Beam Aerials

The idea of a proper beam aerial is probably pie in the sky for most people, but for the lucky few who have the space and the means to put one up it can make an amazing difference to the performance of the station. Not only does a beam have a significant amount of forward gain, but it will also reduce the level of signals which are not in the direction of the aerial. This means that levels of interference will be reduced quite significantly. In fact people who use beams will often be able to copy a station quite clearly when other local stations using lesser aerials will not even be able to hear it.

Most beam aerials for use on the HF bands are made commercially in view of the mechanical problems which would be encountered by the home constructor. Although having said this a number of people have very successfully managed to build their own. If this is done then it is necessary to ensure that it is suitably constructed to withstand the rigours of the weather. It is surprising just how much wind resistance even a simple aerial can generate when the wind rises.

Sometimes these HF beams are monoband aerials but more usually they use traps to enable them to cover the three most popular DX bands: Ten, Fifteen and Twenty metres. The fact that traps are used within the aerial does mean that the overall size is reduced somewhat. Even so these aerials are still quite large.

In an attempt to reduce the size of these beam aerials there are a number of miniaturized beams available. They are somewhat smaller than the other beams and use the rather distinctive capacity hat traps which consist of a circle of wires about 6 to 10 inches long coming out from the elements. Whilst these designs have the advantage of being smaller in size than their trapped counterparts there are some penalties to pay. The first is that they offer a reduced gain over their operating bandwidth. The second is that their bandwidth is markedly smaller. This can be a problem because it means that the aerial can normally operate only over a portion of each of the bands for which it is intended. Beyond this limit the VSWR will rise sharply and the gain will fall off. When considering buying one of these aerials it is worth looking at its bandwidth and judging whether it is satisfactory for the needs of the station.

In order to realise the full potential of a beam it should be mounted clear of all the obstructions around. This fact alone often gives an aerial a significant advantage over another aerial which might have been erected. However to do this will probably mean that a tower or large mast will have to be installed and this can cost more than the aerial itself. This obviously means that installing the aerial becomes a major undertaking, but it means that the aerial system will work well.

**Aerial Tuning Units**

Aerial tuning units (A.T.U.) in one form or another are found in most S.W.L. and transmitting stations. For listening they can be very useful as they can provide an increase in the signal strength from the aerial whilst for transmitting they can ensure that the transmitter output sees a low VSWR and the right impedance. This is particularly important with many of today's all-transistor rigs because either there is a possibility of damage to the P.A. if it does not have any protection or alternatively the protection will reduce the output power quite significantly in some cases.

Although the basic idea of an A.T.U. is to provide a good match between the equipment and the aerial, there are a few small differences in the way they can be used. In some instances they are connected directly to the aerial whilst at other times they may be used at the equipment end of a coaxial feeder.

End fed wire aerials provide one of the most obvious uses for an A.T.U. As they can be virtually any length and they can be used over a whole range of frequencies, the impedance which they present at the feed point can be almost anything. In order to provide a good match to the feeder or equipment an A.T.U. is necessary, but fortunately it can be quite simple in its design. A circuit for one which could be used in this situation is shown in Figure 4.10. Often suitable components can be found from items already in the shack. The

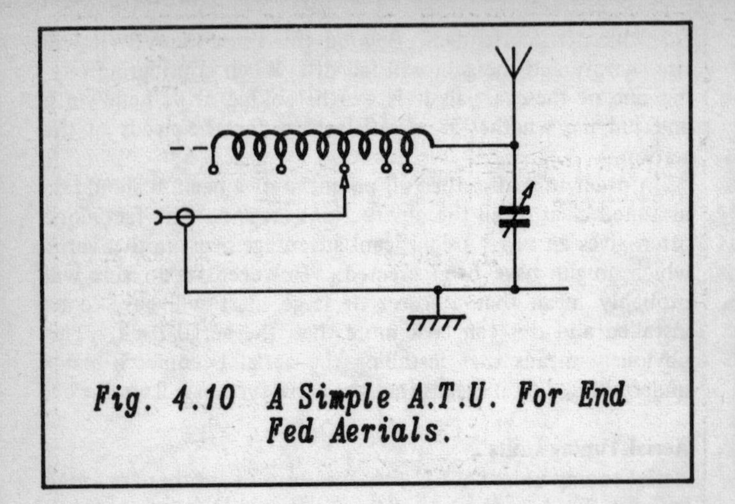

Fig. 4.10 A Simple A.T.U. For End Fed Aerials.

variable capacitor should be up to about 500 pF although this is by no means critical. However, if it is to be used for transmitting it should be a widespaced one. The coil should be an inch or two in diameter and it should have fifty turns or more of 16 s.w.g. wire or thicker, depending upon the bands likely to be used. Taps should be placed along the winding, spaced every turn or so at the beginning, but spaced out more further along its length.

Another instance where an A.T.U. is needed is when an aerial using open wire feeder is used. As many of these aerials will be used over a wide band of frequencies high levels of standing waves will exist on the feeder, and it will present a very poor match to the transmitter (or receiver). In order for the system to operate correctly an A.T.U. is needed to match the system properly. However, the circuit of the tuning unit needed to accomplish this will be somewhat different and it will need to have a balanced output.

Finally an A.T.U. can be used at the equipment end of a coaxial line as shown in Figure 4.11. Although it will not be able to match the aerial itself to the coaxial feeder it is useful here because it will reduce the SWR seen by the transmitter. As a result it can help protect the P.A. of the transmitter from the effects of a high standing wave ratio.

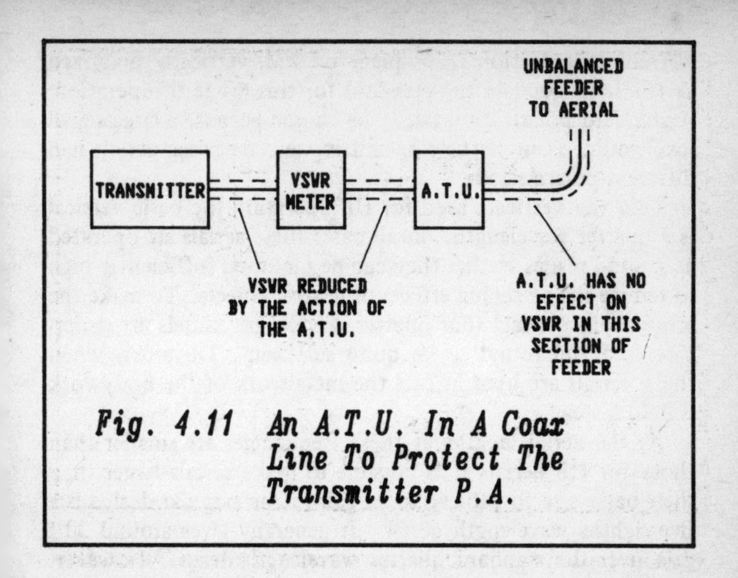


*Fig. 4.11 An A.T.U. In A Coax Line To Protect The Transmitter P.A.*

**Aerials for VHF and UHF**

Aerials for VHF and UHF differ slightly from those used on the lower frequencies. Although the basic principles of operation do not change, the basic sizes of the aerial elements are smaller. This means that aerials with much higher gains can be considered. In fact it is not uncommon for people to use aerials with gains of 15dB or possibly more which would be unheard of on the HF bands. In addition to this the way in which radio is used is slightly different. For example, mobile operation is widespread on these frequencies whereas there are very few mobile stations active on the HF bands. Differences like these mean that there are a number of aerial designs which are commonly used on the VHF and UHF bands which are seldom seen at HF.

**Vertical Aerials for VHF/UHF**

The widespread popularity of mobile operation has given rise to the necessity for omnidirectional aerials. As a result various forms of vertical aerials have become very popular because they are both omnidirectional and easy to make. As virtually all mobile operation takes place on FM, vertically polarized aerials have become the standard for this mode of operation. A standard polarization has to be chosen because a large signal loss would occur if the transmitting and receiving aerials had different polarizations.

Like the verticals used for HF operation the basic vertical is a quarter wavelength. In all cases these aerials are operated as ground planes so that they can be mounted sufficiently high to reduce the screening effects of nearby objects. To make the ground plane itself four quarter wavelength radials are generally used and found to be quite sufficient. Obviously when these aerials are used in cars the metalwork of the bodywork is used instead.

As the aerial lengths at these frequencies are smaller than those for HF aerials it is possible to make aerials larger than their basic size to achieve some gain. One popular design is a five-eighths wavelength aerial. It generally gives around 3dB gain over the standard quarter wavelength design. However, as one might imagine its impedance is not 50 ohms because it is not fed at a point where current is at its maximum and the voltage is a minimum. In order to overcome this problem these aerials have a matching coil in the base.

Another popular design is a collinear. Although a collinear can exist in many forms it has become very popular at frequencies above 100 MHz or so as a vertical. Essentially it consists of a number of half wave elements fed in phase. The effect of this is that the radiation is concentrated at right angles to the axis of the aerial. The gain is dependent upon the number of elements and the more there are then the greater the gain. Using a collinear gains of up to 9dB or so are available, but most of the aerials offer less. These aerials can be quite tall because they have several half wave sections on top of one another. For example, the aerial offering 9 dB gain at 430 MHz is over 3 metres long. In addition to this these aerials are fairly expensive. This is due to the height and the fact that even though it may appear to be a single vertical element from a distance, it actually consists of several sections.

### Horizontal Aerials

Although vertical polarization works very well for FM operation which is essentially local, research performed over a number of years has shown that horizontal polarization is better for long distance operation. This effect is more noticeable above 30 MHz or so because on the HF bands signals reflected by the ionosphere are virtually random in their polarization. When signals are not reflected in this way they tend to keep their polarization better. As a result of these findings most contacts on the traditionally long distance modes are made using horizontal polarization.

Whilst it is possible to use a simple dipole, most VHF and UHF aerials are designed to give significant amounts of gain. In fact when choosing an aerial it will be necessary to decide upon how much gain will be needed. Obviously the more gain the better the contacts that can be made. Against this the higher gain aerials are more expensive and physically larger. In addition to this as the gain of the aerial increases the beam-width reduces calling for more accurate setting of the direction of the aerial. The final decision of the requirements of the aerial is a compromise which will depend on the individual requirements of the station.

There are a number of different types of aerial which can be used. The most popular by far is the conventional Yagi shown in Figure 4.12. It offers several advantages. First of all it can be made to give quite high levels of gain dependent upon the number of elements it has. For example, gains of 10dB or more are quite common. It is also easy to construct making it a relatively cheap form of aerial. Finally it is quite rugged and offers a lower wind resistance than many other forms of aerial with a similar amount of gain.

Another form of aerial which enjoyed a certain amount of popularity was the cubical quad shown in Figure 4.13. This type of aerial offers about 2dB more gain than a Yagi with the same number of elements. However it does have the disadvantage that it is not quite as easy to make which means that it is more expensive than an equivalent Yagi. Also it is not as rugged. These factors have meant that its popularity has declined over the last few years and it is not seen as often.

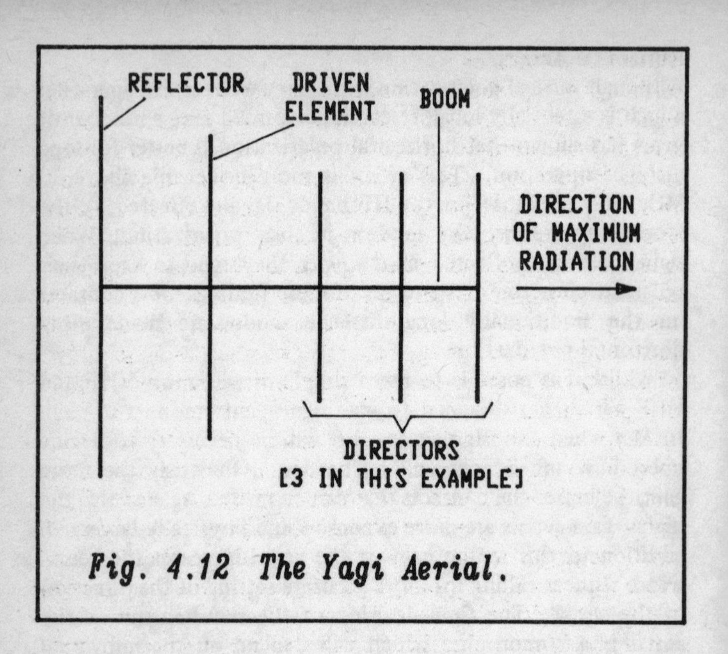


*Fig. 4.12 The Yagi Aerial.*

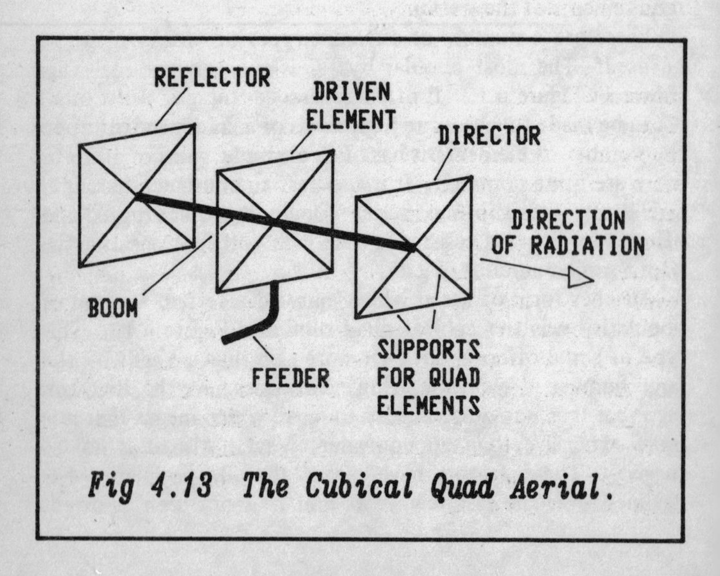


*Fig 4.13 The Cubical Quad Aerial.*

**Other Considerations**
Although it is possible to build aerials for the VHF and UHF bands it is not as usual to do this as it is for aerials for the HF bands. The reason for this is that there is a very good selection of ready-built aerials at very reasonable prices. In addition to this comparatively few people have easy access to the materials and equipment needed to make a robust aerial for these frequencies. If one was to build an aerial for these frequencies the cost saving would probably be quite small. It would also take up a reasonable amount of time, especially if it was to be made to withstand the rigours of the weather. However, if the aerial is to be used internally then home construction is more of a possibility.

When using a directional aerial, it is necessary to have some method of rotating the aerial. The most common way of doing this is to have an electrical rotator. However, rotators are fairly costly and the expense of this needs to be borne in mind when planning the system out. Often they can cost more than the aerial itself. Also the size of the rotator needs to be tailored to the size of the aerial, and any future aerials which are likely to be added to the system. Dependent upon this the correct size of rotator can be bought.

**Choice of Feeder**
The choice of the feeder to be used in a station can be very important. After all there is little point in spending a lot of time and money in erecting a good aerial and then losing a lot of the signal by employing poor feeder. Conversely as feeder can be very expensive, it is not wise to spend a lot of money on very low loss feeder if it will not reap any worthwhile benefits. Generally for short feeder runs using low frequencies and low power levels a cheaper grade feeder need only be used. However, for higher frequencies where the feeder losses rise, and for longer runs, then a more expensive cable is a necessity.

Cable comes in a wide variety of different types, each having its own characteristics in terms of dimensions, loss and impedance. A summary of many of the more common types is given in Figure 4.14.

| Type | Characteristic Impedance (ohms) | Outside Diameter (mm) | Velocity Factor | Attenuation (dB/10 metres) | | Comments |
|---|---|---|---|---|---|---|
| | | | | @ 100 MHz | @ 1000 MHz | |
| RG5/U | 52.5 | 8.4 | 0.66 | 1.00 | 3.8 | |
| RG9/U | 51.0 | 10.7 | 0.66 | 0.66 | 2.4 | |
| RG10A/U | 50 | 12.1 | 0.66 | 0.66 | 2.6 | |
| RG11A/U | 75 | 10.3 | 0.66 | 0.76 | 2.6 | |
| RG12A/U | 75 | 12.1 | 0.66 | 0.76 | 2.6 | |
| RG20A/U | 50 | 30.4 | 0.66 | 0.22 | 1.2 | |
| RG58C/U | 50 | 5.0 | 0.66 | 1.8 | 7.6 | |
| RG59B/U | 75 | 6.1 | 0.66 | 1.2 | 4.6 | |
| RG62A/U | 93 | 6.1 | 0.84 | 0.9 | 2.8 | |
| RG213/U | 50 | 10.3 | 0.66 | 0.62 | 2.6 | Polythene dielectric |
| RG214/U | 50 | 10.8 | 0.66 | 0.76 | 2.9 | Double screened. Silver plated copper wire. |
| RG223/U | 50 | 5.5 | 0.66 | 1.58 | 5.4 | |
| RG316 | 50 | 2.6 | | 3.8 | 11.8 | Silver plated steel entre conductor. Silver plated copper braid. |
| UR43 | 50 | 5.0 | 0.66 | 1.3 | 4.46 | |
| UR57 | 75 | 10.2 | 0.66 | 0.63 | 2.3 | Similar to RG11A/U |
| UR67 | 50 | 10.3 | 0.66 | 0.66 | 2.52 | Similar to RG213/U |
| UR74 | 51 | 22.1 | 0.66 | 0.33 | 1.4 | |
| UR76 | 51 | 5.0 | 0.66 | 1.7 | 7.3 | Similar to RG58C/U |
| UR77 | 75 | 22.1 | 0.66 | 0.33 | 1.4 | |
| UR79 | 50 | 21.7 | 0.96 | 0.17 | 0.6 | |
| UR90 | 75 | 6.1 | 0.66 | 1.2 | 4.1 | Similar to RG59B/U |
| *Standard TV Coax | 75 | 5.1 | 0.66 | 1.1 | 4.0 | |
| *Low Loss TV Coax | 75 | 7.25 | 0.86 | 0.75 | 2.6 | Semi-air spaced. |

*These cables are not standardized. Figures given are typical.

*Figure 4.14 Coax Cable Specifications*

**Further Reading**

"25 Simple Amateur Band Aerials" (BP125) by E.M. Noll. Publisher: Bernard Babani (publishing) Ltd
ISBN 0 85934 100 3

"An Introduction to Antenna Theory" (BP198) by H.C. Wright. Publisher: Bernard Babani (publishing) Ltd
ISBN 0 85934 173 9

"HF Antennas For All Locations" by L.A. Moxon G6XN
Publisher: Radio Society of Great Britain
ISBN 0 900612 57 6

# Chapter 5

# CONSTRUCTION

In any station there will be a certain level of construction needed. In some instances it may only be attending to the various repairs which are needed from time to time. It may be that a lead has to be made up. Then at other times a full constructional project can be undertaken. In fact, whatever the size of the job the same constructional skills are needed but obviously to a greater or lesser degree.

When building a project one of the first things to be considered is whether the size of the project is too large. Unfortunately many radio shacks are littered with a number of half finished projects. These projects could remain incomplete for a number of reasons. However, one of the most common reasons must be that they are too large and either the enthusiasm runs out or they become too costly and complicated.

When starting a project a number of points have to be considered. Everything from the size of the project to the type of board to be used as well as the case for it. However, one of the fundamental aspects of construction is soldering.

## Soldering

Good soldering is at the heart of electronic construction. Whether used for amateur projects or in professional equipment it is equally important. It is also a skill which has to be learned. Fortunately this is not difficult to do, but it requires a little care and patience, because a rushed job will always show in the neatness and the quality of the joints which are made.

When considering whether to start some construction the first requirement will be a suitable soldering iron. For most work a light iron dissipating about 15 watts and no thermostatic temperature control will be quite adequate. However, for the more ambitious constructor the luxury of a more expensive iron with full temperature control can be an advantage. Either type of iron will be suitable for normal constructional projects and general jobs around the shack. Occasionally there will be a need for a more powerful iron. For example when installing an aerial using thick wire, or possibly fitting in an earthing system. Whilst many amateurs will have a large iron for jobs like these, many will not and manage to improvise quite happily or just borrow an iron when necessary.

Having obtained a suitable iron it is also necessary to use the right form of solder. All electronic components stockists will have supplies of the right sort of solder. It should have a flux core. This is necessary as the flux serves to remove any oxide around the joint, helping to ensure a good connection. Although plumbers solder with a separate flux is widely available at hardware shops, this is not suitable for electrical work and should not be used.

When making a joint both surfaces should be clean. If not they should be carefully cleaned to ensure a good joint. When making a joint it is necessary to apply the iron and the solder to the joint at the same time. The solder should not be applied to the iron and then transferred to the joint as this is a recipe for a bad connection. However, it is a good idea to have just a small amount of solder on the iron as this will help the heat to be transferred to the joint.

To make the best joints the soldering should be done reasonably quickly. If the solder is kept hot for too long by the iron then the flux will become used up allowing oxidation and this will lead to what is called a dry joint. Dry joints look frosted rather than the shiny finish of a good one, and they give a poor electrical connection which can often be intermittent.

It is also necessary not to use too much solder. Enough should be used to make a good joint, but it should not form a large blob around the components. After a little practice it will become obvious how much should be used.

After a few joints have been made it will be found that the soldering iron bit will become dirty and it may have excess solder on it. This can be removed by quickly wiping the bit on some paper or a wet sponge. In fact many soldering iron stands have a special place moulded into the plastic for a sponge.

Having completed the soldering some dark flux will remain round the joints. This may look a little unsightly on a board. It can be removed if necessary, although with today's fluxes it does no harm to leave it. If it is to be removed then it may be dissolved in alcohol.

## Forms of Construction

There are many different methods which can be used when building up a circuit for a project. At the bottom of the scale there is the proverbial rat's nest where components are just soldered to one another and gradually the circuit evolves. Whilst this may just about be alright for the odd experimental circuit, it is most unsatisfactory if it is to be permanent. In fact even for experimental purposes a tag strip form of construction would be better. At the other end of the scale are the professionally finished printed circuit boards. In addition to these methods there are a number of intermediate methods which can be used. One of these is to use small pins and wire on a matrix board, or strip board could be used. Each method of construction has its own advantages and can be used to good effect at different times.

## Stripboard

Stripboard, or to give it one of its trade names Veroboard, is a popular form of construction for low frequency projects. For example, it is very good for some audio or logic types of circuit. The stripboard shown in Figure 5.1 makes a good base for building up a circuit. Essentially it consists of a matrix board having holes at regular intervals (normally 0.1 inch spacing). On one side of the board there are copper strips as shown which can be used for connecting the different components as required. The components are inserted through the board and soldered. This makes an easy way of connecting the components.

If a strip needs to be cut this can be accomplished quite easily using a specially designed hand tool called a spot face cutter. Alternatively it is possible to use a drill bit. Both tools act by drilling through the copper strip and a small amount of the board to ensure that there is no connection between the two sections. When doing this it is absolutely necessary to be

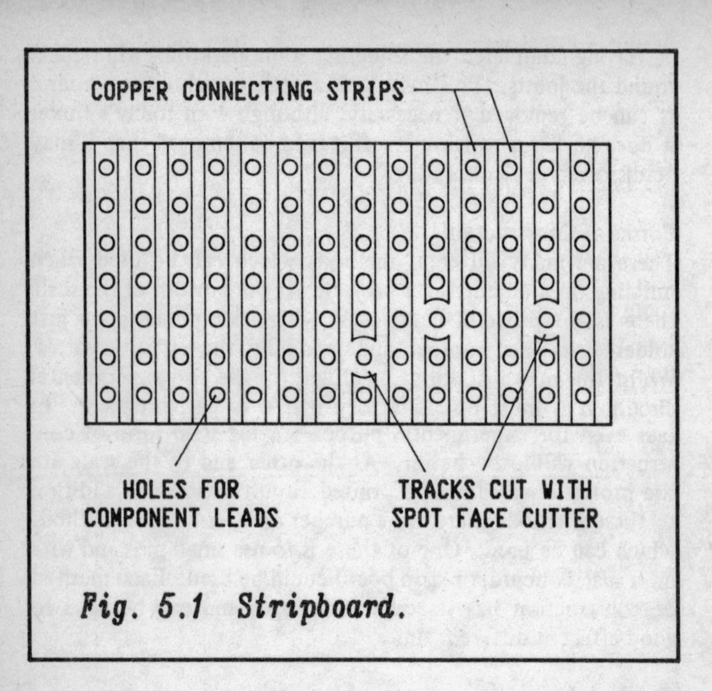


*Fig. 5.1 Stripboard.*

careful that connection between the two sections is broken. It is very easy to leave a minute whisker of copper between the two sections that is very difficult to see. Another problem which can arise is that these small whiskers can also short between adjacent tracks. Apart from taking care when actually using the spot face cutters, it is a wise precaution to run a small screwdriver up and down the gaps between adjacent tracks and across the track cuts. This will help ensure that there are no whiskers of copper to give short circuits between adjacent parts of the circuit.

Stripboard has several advantages. It does not require much preparation for it to be used, unlike printed circuit boards which have to be designed and made. In addition to this it is relatively cheap and it is readily available. However, it does have some disadvantages. It can be difficult to lay a circuit out in a logical fashion which is easy to follow for fault finding. In addition to this the difficulties of laying out a circuit mean that it is not ideal for any RF projects. Even so if it is used for low frequency circuits it does prove to be a satisfactory and convenient base to use for building circuits.

**Pin and Wire**

One form of construction which can be tailored a little better to RF projects is the use of plain matrix boards with pins and wire. The matrix board is very similar to the stripboard having holes usually on a 0.1 inch matrix. However, components are mounted through the board and their leads are connected to pins as shown in Figure 5.2. Other components can be connected to the pin, or a wire can be used to route the connection to other points on the board as required.

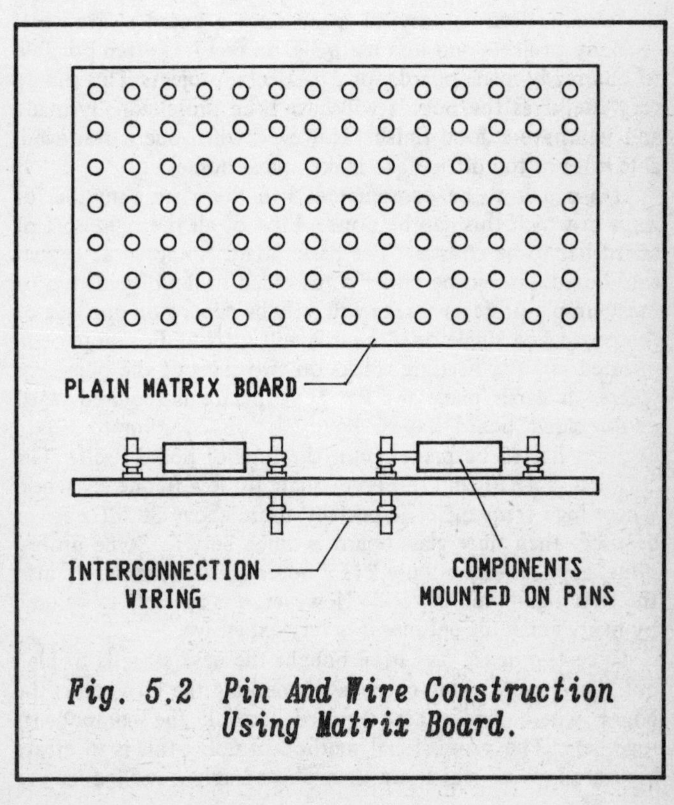


*Fig. 5.2 Pin And Wire Construction Using Matrix Board.*

Although this method of construction does take longer to build it does afford much more flexibility than stripboard. This means that sensitive leads can be kept short and the circuit can be arranged more easily as required. Also earth leads can be made thicker if needed. Even so it does have its shortcomings and is certainly not ideal for use above 30 MHz.

**Printed Circuits**

Printed circuit boards (pcbs) are widely used as the basis on which to build a wide variety of circuits from the amateur to the professional. They have many advantages. They are neat, robust and once a board has been designed it can be reproduced and the circuit can be reliably repeated. In addition to this the circuit itself is easy to make once a board is available.

Many projects and kits use pcbs. In fact it is often possible to buy ready-made boards for a particular project. This can be very useful as the boards will have been professionally made and will have a good finish. However, when one is not available it is not too difficult to make one at home.

If a pcb is to be constructed then there are a number of ways in which this can be done. First of all the right sort of board has to be chosen. The basic board is known as copper clad board. It can be either single-sided or double-sided. For most simple projects tracks will only be needed on one side of the board and single-sided boards will suffice. For more complicated circuits needing tracks on two sides of the board, or where an earth plane for RF performance is required, then double-sided board has to be used. In addition to this a decision has to be made about the type of board itself. The cheapest is SRBP and this is adequate for low frequency work. Where high frequencies, especially those above 30 MHz are to be used, then fibre glass board is much better. Some professional applications will use PTFE board for frequencies up into the UHF bands and beyond. However, it is unlikely to be used by many amateurs because it is very expensive.

Once the board has been bought the next stage is to plan out the circuit. This consists of marking the areas onto the board where tracks are required to link the components together. The professional method of doing this is to create accurate layouts which are then photographed and reduced to the correct size. This image is then projected onto the board covered in a substance called photoresist. The board is then "developed" to leave only the areas which need tracks covered by the protective photoresist. This method is not normally viable for the amateur. For one off boards which are likely to be experimental it is much better to mark the tracks directly onto the board. Almost any etch resist material can be used. Special transfers are available, as are fibre tip pens with etch resist ink, and even many types of paint can be used. Essentially almost anything which prevents the etching solution from attacking the copper can be used. However, it must be possible to remove it easily once the etching process has been completed.

Once the layout is complete the next stage is to etch the board itself. There are several chemicals which can be used for this, but the most common one is ferric chloride. The solution should be made up and used as directed on the packet. The board can be placed into the solution and left until all the unwanted copper has been removed. It should not be left any longer than necessary otherwise the etch solution will tend to creep under the resist material making an untidy finish.

When using any etch material it must be remembered that the solution is corrosive and the greatest care MUST be taken. Accordingly, rubber gloves should be worn and if any comes into contact with the skin it should be washed off immediately. Great care should also be taken to ensure it does not come into contact with the eyes. In addition to this, ferric chloride will leave a nasty stain if it is not removed quickly from anywhere it may have been spilled or splashed.

Once the board has been etched it should be washed to remove any etch solution and then the etch resist can be removed. Having done this the holes for the components can be drilled. If double-sided board has been used with the second side left intact as an earth plane, then the earth plane has to be cleared around the holes to prevent the leads shorting to earth. This can be easily accomplished after the component holes have been drilled by countersinking the holes slightly from the earth plane side.

With all the holes drilled the board is ready for the addition of the components, and completion of the circuit. Full details of making printed circuit boards are given in the book mentioned at the end of this chapter.

**Unit Wiring**

Once a board has been built it will be necessary to wire it into the final unit. This requires as much, if not more, skill than making up the board itself if the final appearance is to be neat and tidy. This is best accomplished if the wires are run together and then when all the connections have been made the wires are tied together to form what is called a wiring loom.

Most electronics components suppliers will have stocks of special lacing cord for this purpose. It generally comes in two types: a plastic covered sort and a waxed linen variety. Of the two the waxed linen cord is the best to use as it does not come undone as easily.

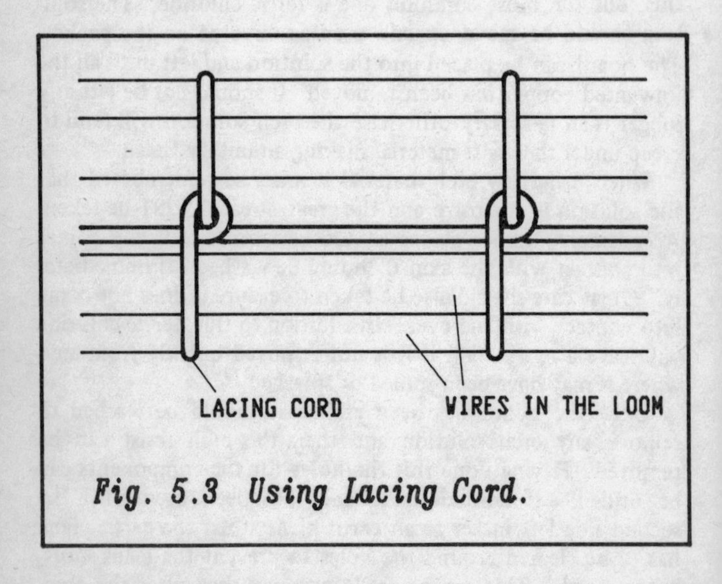


*Fig. 5.3 Using Lacing Cord.*

The wires should be carefully laced together. The cord should be run along the length of the loom and tied at intervals of an inch or so as shown in Figure 5.3. The exact spacing is not critical but will depend to some extent on the number of wires in the loom and the control which is needed for the wires.

**Further Reading**
"How to Design and Make Your Own P.C.B.s" (BP121) by R. A. Penfold. Publisher: Bernard Babani (publishing) Ltd ISBN 0 85934 096 1

## Chapter 6

# TEST EQUIPMENT AND TESTING

A certain amount of test equipment is essential in any amateur radio station. When even small problems arise then it is necessary to have some form of test equipment, however simple, to help locate the problem. Obviously if a certain amount of construction is envisaged, then it helps to have a wider variety of equipment to enable the circuits to be completed. All too often projects seem to be left just as they are before reaching completion, and all because they do not work quite as they should and the fault cannot be located properly.

Fortunately the average radio station does not need a whole set of expensive equipment to be bought. A little ingenuity and know-how can enable a lot of problems to be solved with only a minimal selection of test equipment. However, some items are almost invaluable and others can be made to perform a very wide variety of functions which might otherwise require a number of different items to be used.

### The Multimeter

The most useful piece of test equipment in any radio station must be a test meter. In fact a basic meter which measures current, voltage and resistance is a necessity. Even if servicing of the more complicated equipment is never likely to be undertaken, some form of meter is still needed for performing some simple checks when setting up and maintaining a station. Then if constructional projects are started it becomes even more important.

The choice of the meter is quite important. Obviously the more money that is spent then the better the meter will be. However, it is necessary to consider exactly what is needed. Most measurements are likely to need only an indication of the approximate value. For example it might be necessary to find out if the voltage is getting to a point on the circuit, or possibly whether a fuse is open circuit or not. In both these cases the actual value of the voltage or resistance is of little value. These are the most common forms of measurement and really only need a fairly basic meter. However there will always be the time when an accurate indication will be needed and this is where the better meter is very useful.

The other factor to be considered is the physical robustness of the meter. When making measurements meters will invariably be perched on odd corners of the bench and eventually the time comes when they get dropped and even if one is careful they will tend to be knocked around to a certain extent. With this in mind a more robust and expensive meter might be a good investment.

The other aspect to consider is whether to buy an analogue or digital meter. Today there is a very wide choice of meters for sale, from the very cheapest analogue ones to the more expensive ones which include the digital ones.

Analogue meters have the advantage that the trend or approximate value of a reading can be seen very easily at a glance. However they are not so easy to use for accurate readings. Care has to be taken over items like parallax errors. Even then the best analogue test meters will only achieve accuracies of around 1%, which it must be said is more than accurate enough for all but the most exacting purposes.

Digital meters are capable of making much more accurate measurements. A reasonable digital meter will be capable of achieving accuracies of better than 0.5% on most of the ranges and the better meters will be even more accurate than this. They can also have higher impedances on the voltage ranges which mean that they hardly make any difference to the circuit when they are making a measurement. In addition to this a lot of the new digital meters include a number of extra facilities like transistor, capacitance or even frequency measurements at little extra cost. As one might expect digital meters are often more expensive than at least the cheaper analogue ones. But with the prices of the digital ones falling and the analogue ones remaining static, the price differential is becoming smaller.

**Metering In Circuit**

Whatever meter is applied to a circuit it will have some effect, however small. In many instances the effect will be so small as to be absolutely negligible, but in other cases there will be some effect which has to be taken into account.

Voltage measurements form the vast majority of those which are made with a test meter. The possibility of any test meter measurements altering the operation of the circuit will often depend upon the meter being used where DC or low frequencies are concerned. Usually the effect a meter has on any circuit will depend upon its resistance. It will act as a resistor across the two points being measured. If a digital meter is used then its resistance will normally be very high; usually of the order of several megohms. However, where analogue meters are concerned it is a little more complicated. The meter sensitivity will be specified in terms of a certain number of ohms per volt. Essentially this means that the meter will have a resistance of that number of ohms for every volt of the particular range in use. For example, a meter having a sensitivity of 20k ohms per volt and operating on a 10-volt range will have a resistance of 200k ohms. From a knowledge of this resistance and the circuit in question it is possible to judge whether the measurement will affect the circuit operation.

Usually the effect of a meter on the DC conditions is negligible. However, the same is very seldom true when radio frequencies are involved. Even a small length of wire, let alone a test lead, can have a dramatic effect on the operation of a circuit. This is particularly true of tuned circuits where even the addition of a few picofarads can totally detune the circuit. There are two ways of overcoming this problem. The first is not to monitor any points in the circuit which are carrying any RF, but this will obviously restrict the measurements which can be made. However, when measurements have to be made on a point where RF is present then the effect of the test leads can be reduced to an extent by adding a resistor onto the end of the test lead as shown in Figure 6.1. The exact value of the resistor will not be critical, but it should be as high as possible consistent with not affecting the accuracy of the meter. A value of 10k ohms for an analogue meter might be satisfactory, whilst it is quite possible to increase this to 100k ohms or more for a digital meter. When this resistor is added the end of the resistor in contact with the circuit should have as short a lead as possible. This will help

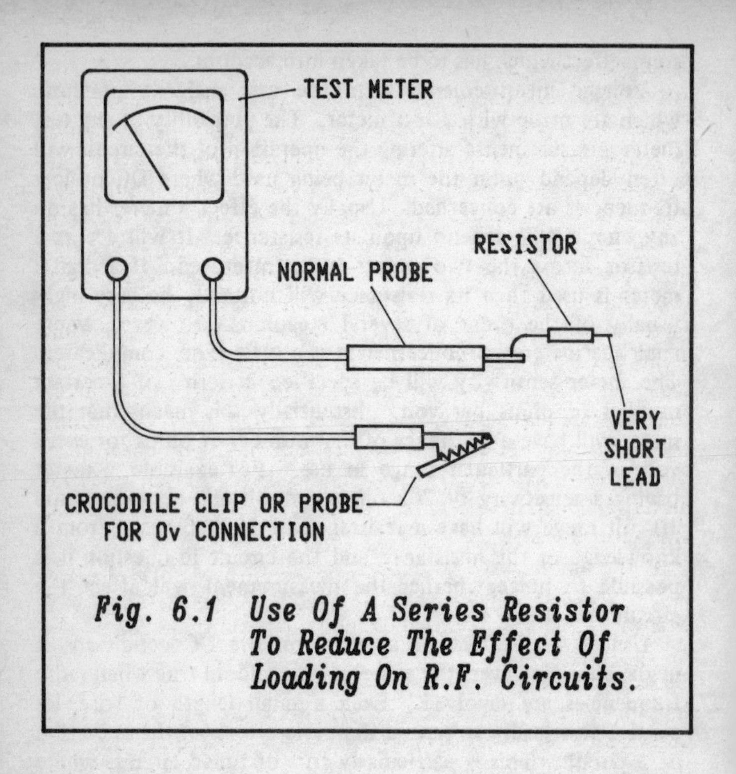


*Fig. 6.1 Use Of A Series Resistor To Reduce The Effect Of Loading On R.F. Circuits.*

reduce the value of any stray capacitance. Even so the presence of the resistor may well affect the circuit slightly and this should be borne in mind when taking measurements.

Current measurements are not usually a problem for circuits which do not carry any RF. The main problem would be the resistance of the meter and under normal circumstances its resistance is so low as to be negligible.

However, if the circuit is carrying RF then it is quite possible that there may be a few problems which are encountered. Even if the line being measured is supposedly decoupled and should not be carrying any signals, some may leak onto it. This would in turn be passed onto the long meter leads and give an increased possibility of unwanted feedback or other effects.

**A Simple Diode Test**

Although many digital meters these days have a diode test facility built into them, analogue ones do not. Fortunately it is quite easy to perform a quick and easy confidence check by using the ohms range on the meter. The test simply relies on the fact that diodes will conduct in one direction and not in the other.

The first stage is to check that the diode conducts in the forward direction. To do this connect the cathode to the meter terminal marked +ve and the anode to the other terminal as in Figure 6.2(a). The meter should deflect across the scale. It is not possible to say what the reading should be as this depends upon the meter and the diode. Having done this the connections should be reversed and the diode checked in the opposite direction as in Figure 6.2(b). This time the meter should not deflect at all or at least only a very small amount if the diode is germanium.

If the diode conducts only in one direction then it is likely to be properly functional. Any fault is likely to be quite visible, i.e., it will conduct in both directions or not at all.

**Transistor Test**

It is possible to extend the diode test quite easily to give a quick and easy test for a transistor. Although it will not test the performance of the transistor it is normally quite adequate for most test and repair jobs because it is comparatively rare to find a transistor where the device characteristics have changed so that it will not operate properly. Instead it is more usual to find that a catastrophic failure has occurred.

The simple test is based around the fact that a transistor appears like two back-to-back diodes as shown in Figure 6.3. This means that testing the transistor is simply a matter of checking that both diodes are intact. This is done by connecting one lead of the meter to the base and checking between the collector and then the emitter. Then the leads should be reversed and the test repeated. With the one lead of the meter connected to the base there should be a low value of resistance showing that the diode is conducting between the collector and then the emitter. Then in the other direction the resistance measured should be very high.

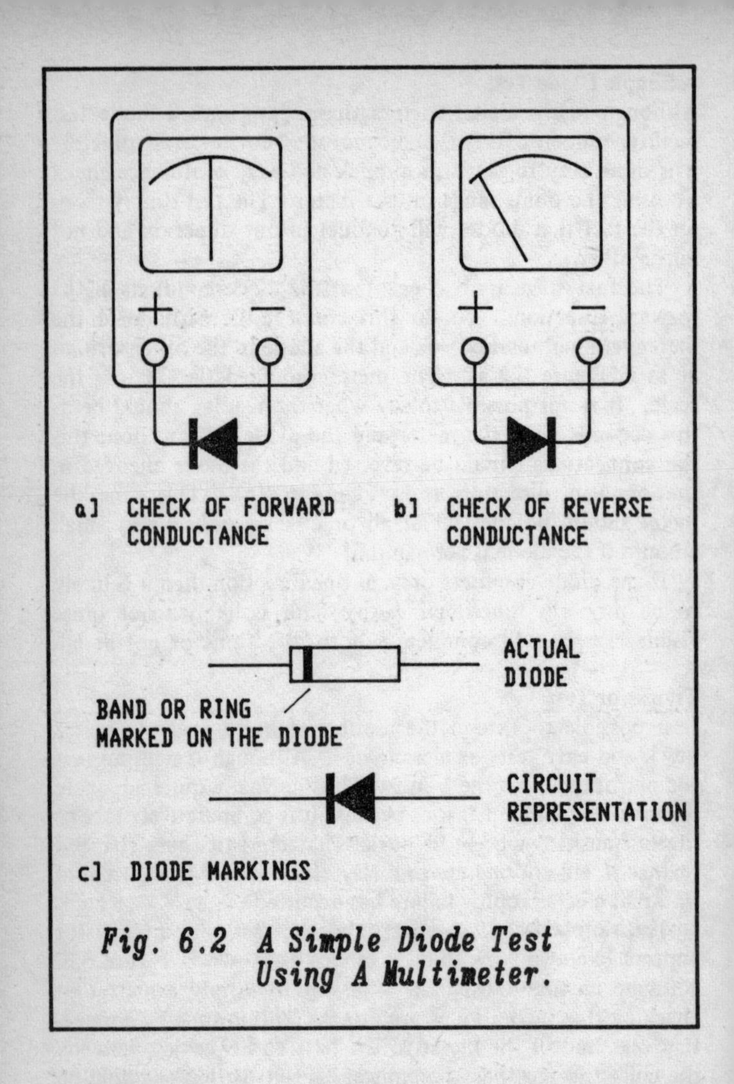

*Fig. 6.2 A Simple Diode Test Using A Multimeter.*

One further check is needed. The resistance between the collector and emitter should be measured. This should be very high. However, if the transistor is faulty it may show a low value of resistance even though it may appear that both the base emitter and base collector junctions are intact.

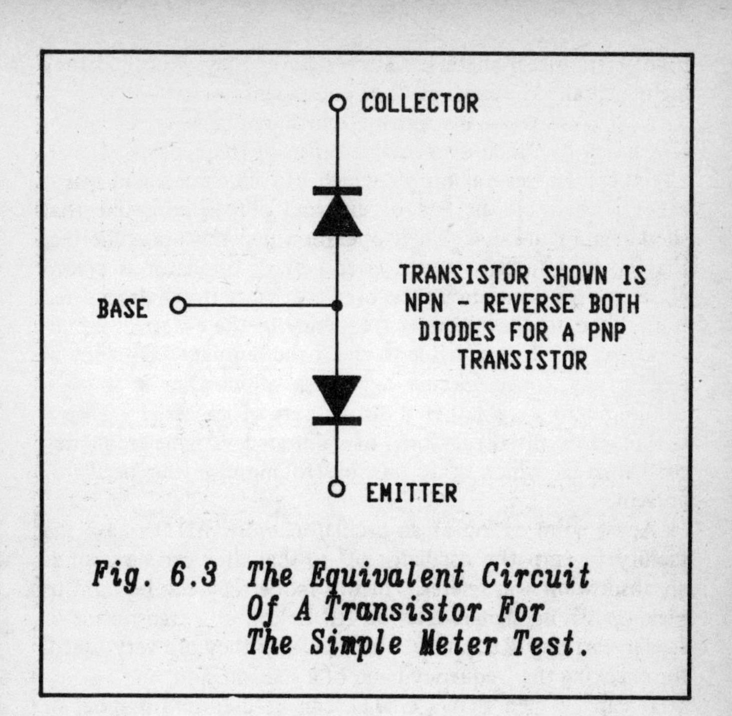

*Fig. 6.3 The Equivalent Circuit Of A Transistor For The Simple Meter Test.*

**G.D.O.**

The G.D.O. must be one of the most versatile instruments available for the radio enthusiast. In its oscillator or dip mode it can measure the resonance of tuned circuits. In its own right this can be very important, but when it is used with a little ingenuity it is possible to use it to make a very wide range of measurements around the amateur shack. Then with the meter in its passive wavemeter mode it can make further measurements adding to the value of this instrument.

G.D.O.s are called a variety of different names depending upon the type of amplifying device used in them. Early meters using valves were all called G.D.O.s (grid dip oscillators) whereas later meters using FETs altered the name slightly to gate dip oscillator. However, they may even be called FET dip oscillators and there may be other names as well for ordinary bipolar transistor meters. But whatever they are called they are essentially the same piece of equipment.

A G.D.O. is an instrument which contains an oscillator which can be tuned over a wide range of frequencies. Generally there are several ranges which can be chosen and this is accomplished by the use of external plug-in coils like that shown in Figure 6.4. Their operation depends upon the fact that when a tank or tuned circuit of an oscillator is placed close to a resonant circuit the oscillator current will drop when it is tuned to the resonant frequency of the external circuit. By doing this it is possible to check the resonant frequency of almost any tuned circuit regardless of whether it is on a circuit board or whether it forms part of an aerial. Thus a G.D.O. is essentially a form of calibrated variable frequency oscillator in which it is possible to monitor the oscillator current.

Apart from acting as an oscillator, most G.D.O.s have the facility to turn the oscillator off so that they can be used as an absorption wavemeter. In this mode they can be used to pick up strong signals like the RF field near a transmitter or feeder carrying RF power. In this form they are very useful for checking the frequency band of a transmission.

It can be seen that a G.D.O. can be used in a number of different ways. By using a little ingenuity they can be used to perform a great variety of measurements which are very useful when setting up and experimenting with aerials.

In view of the rather specialized nature of G.D.O.s they are not always available from the normal electronic component and equipment stockists. In cases where there are problems in locating them from the normal outlets, then it is worth trying a local amateur radio dealer. Even if they do not have one themselves then they will almost certainly be able to advise where to obtain one.

**Measuring the Resonant Frequency of a Tuned Circuit**

The basic function of a G.D.O. is to measure the resonant frequency of a tuned circuit. In its basic form this is done by coupling the coil of the meter to the tuned circuit under test. As the frequency of the G.D.O. is swept across its frequency range it is found that the current drawn by the oscillator drops

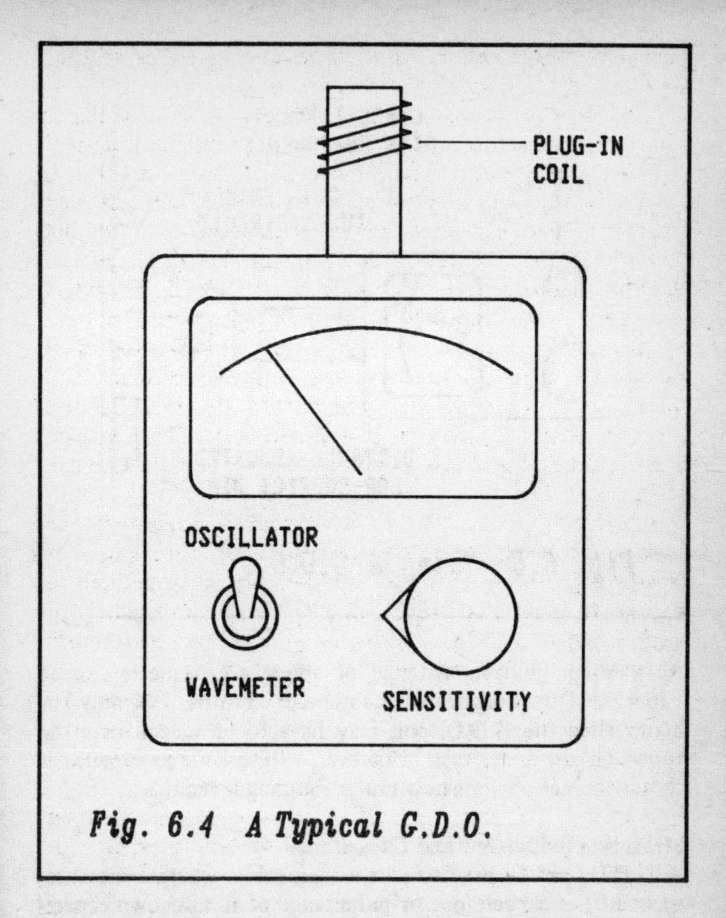

*Fig. 6.4 A Typical G.D.O.*

when the oscillator is tuned to the resonant frequency of the tuned circuit under test.

In practical terms the easiest way of coupling the G.D.O. coil to the tuned circuit under test is simply to place the G.D.O. coil close to the coil of the other tuned circuit as shown in Figure 6.5. Although the actual distance is not critical the greatest distance consistent with a readable dip is best. This is because if the coupling is too tight then the accuracy of the measurement is reduced because the G.D.O. calibration will be altered slightly by the effects of the circuit under test.

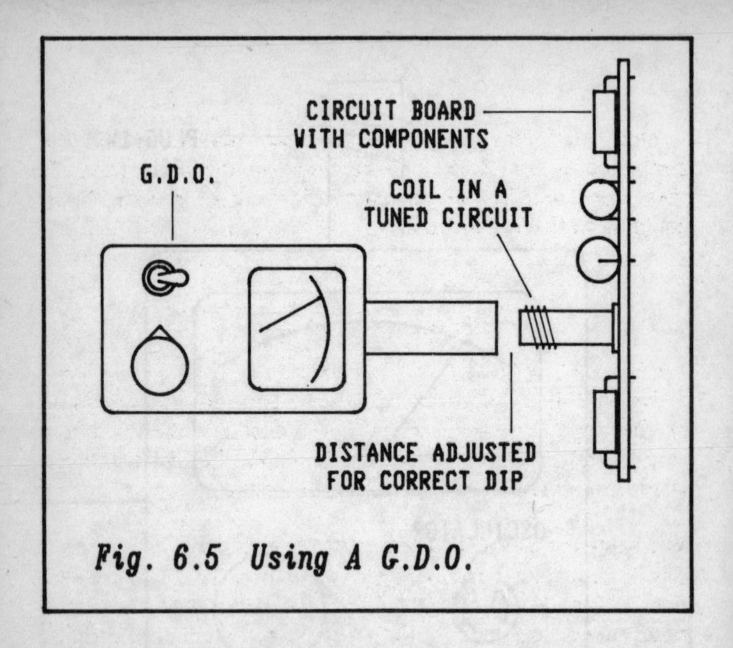

*Fig. 6.5 Using A G.D.O.*

As a rough guide a distance of about a centimetre can be expected if the coil has a large number of turns. If it only has a few then the G.D.O. coil may have to be placed over the tuned circuit under test. However, with a little experience it soon becomes obvious how much coupling is required.

**Measuring Inductance and Capacitance**

A G.D.O. can be used to give a reasonably accurate measurement of the capacitance or inductance of an unknown capacitor or inductor. In order to do this a resonant circuit needs to be made up of an inductor and a capacitor, one of which has a known value. Obviously the more accurate the value of the known item is then the more accurate the measurement will be. In addition to this a little judgement needs to be made when choosing the known component. It is no good placing a large capacitor with a very small inductor because the self inductance of the capacitor will mask out the effect of the inductor. Similarly the same is true with very small capacitors.

Once the resonant circuit has been made up then the resonance of the circuit can be found in the usual way. It then only remains to calculate the value of the unknown component from the formula:

$$f_{dip} = \frac{1}{2\pi\sqrt{(LC)}}$$

Whilst this method may not be as accurate as using some of the new digital capacitance or inductance meters it is a very useful method to have to hand in the average shack where not all the latest instruments are available.

**Measuring the Resonant Frequency of an Aerial**

This is probably the most obvious use for a G.D.O. in connection with an aerial. However, it is not necessarily one of the easiest measurements to make as there are a number of pitfalls.

In common with other measurements for measuring the resonant frequency of a tuned circuit, the basic idea is to couple the coil of the meter to the circuit under test. When the meter is tuned to the resonant frequency of the aerial then the meter current will dip. The centre of the dip indicates the resonant frequency of the aerial.

When performing this measurement it is best to perform it at the aerial itself and not via a feeder. Whilst performing it via a feeder may seem perfectly in order it is found that the feeder will introduce a number of spurious dips and it may be difficult to identify the correct response.

When checking the aerial some way of coupling the meter to the aerial must be found. For an aerial in the HF section of the spectrum it is possible to take a loop of two or three turns of wire from the feed point of the aerial and loop this over the coil of the G.D.O. It may even be possible to use this method having a single turn loop at the low end of the VHF portion of the spectrum, but as the frequency rises this method may introduce some inaccuracies. If sufficient

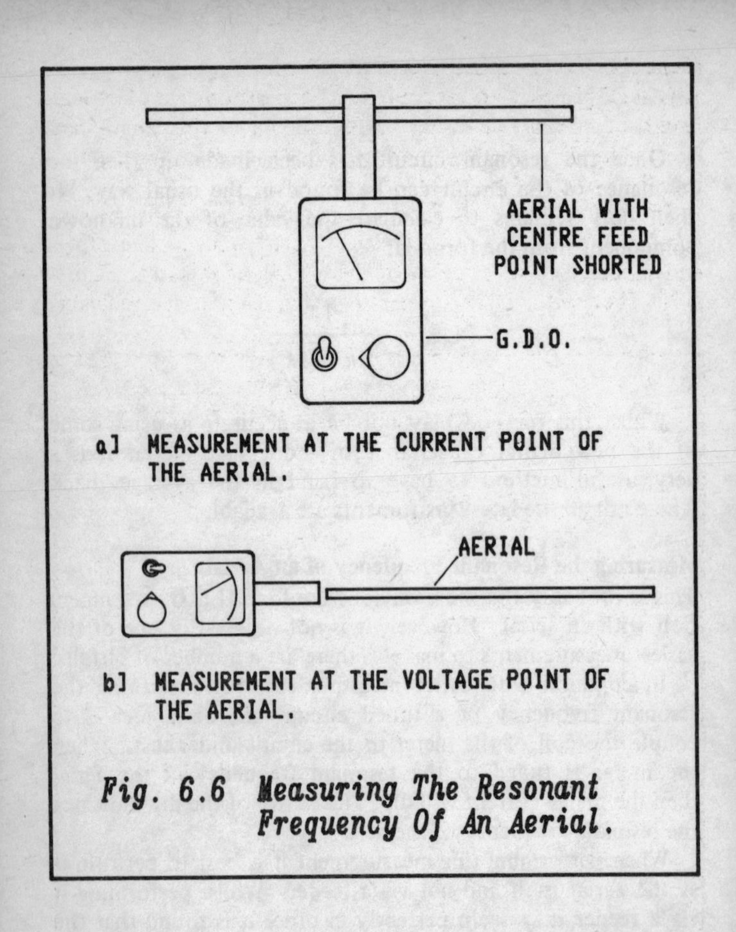


*Fig. 6.6 Measuring The Resonant Frequency Of An Aerial.*

coupling can be obtained the best way is to short out the feed point and place the coil as close as possible to the aerial as shown in Figure 6.6. Using the method shown in Figure 6.6(a) the best dip will be obtained at a current maximum, i.e., at the feed point of most aerials. If the method of Figure 6.6(b) is used then the best dip will be obtained at a point of voltage maximum. This can always be found at the end of an aerial.

**Measuring the Electrical Length of a Feeder**

A G.D.O. provides an easy method of measuring the electrical length of a length of feeder. A knowledge of this length can be of value in a number of applications especially if the aerial installation is used for transmitting.

In order to make the measurement the feeder must be disconnected from the aerial and left open circuit. The other end should then be coupled to the G.D.O. as shown in Figure 6.7. Then with the meter in its oscillator mode it should be

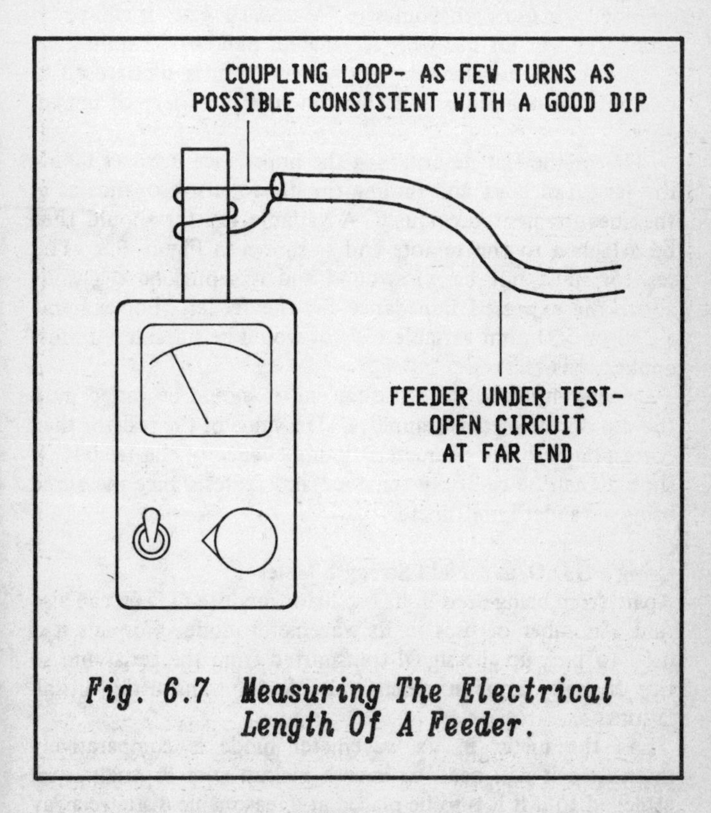


*Fig. 6.7 Measuring The Electrical Length Of A Feeder.*

tuned from its lowest frequency upwards until a dip is noted. This frequency should be noted as it is the primary resonant frequency. However, it is wise to check this by tuning further up in frequency to the next few dips. These are harmonics of the fundamental resonance and they should be at multiples of the frequency of the first dip. If all is correct then the frequency of the first dip corresponds to a quarter wavelength.

**Feeder Impedance**

It is also possible to measure the impedance of a length of feeder. This can be very useful if a length of unknown coax is to hand. Usually it will be 75 ohms if it was originally intended for use with domestic TV or VHF FM. It will be 50 ohms if it was for use with an amateur radio or CB station, or a commercial installation. However, coax is often used in computer installations and this can have a variety of impedances.

The method of determining the impedance involves taking the length of coax and finding the dip for its resonance as in the measurement previously. A variable resistor should then be attached to the remote end as shown in Figure 6.8. This resistor must not be wirewound and it should have a value above the expected impedance for the feeder. For example a 250 or 500 ohm variable resistor would be suitable for most applications.

Having attached it to the far end it should be varied until the dip on the meter disappears. The value of the resistor then corresponds to the characteristic impedance of the feeder. It should then be carefully removed and its resistance measured using a standard multimeter.

**Using a G.D.O. as a Field Strength Meter**

Apart from being used in its oscillator mode, a G.D.O. can also find a number of uses in its wavemeter mode. For this it is used to pick up the signal transmitted from the aerial and so this is really a measurement suitable for transmitting installations.

As the meter in its wavemeter mode is comparatively insensitive it will need to have a pickup wire or small aerial attached to it if it is to be placed at a reasonable distance away from the aerial. This can be set up as shown in Figure 6.9. The best performance will be obtained if the pickup wire is approximately a quarter of a wavelength. Then the meter can

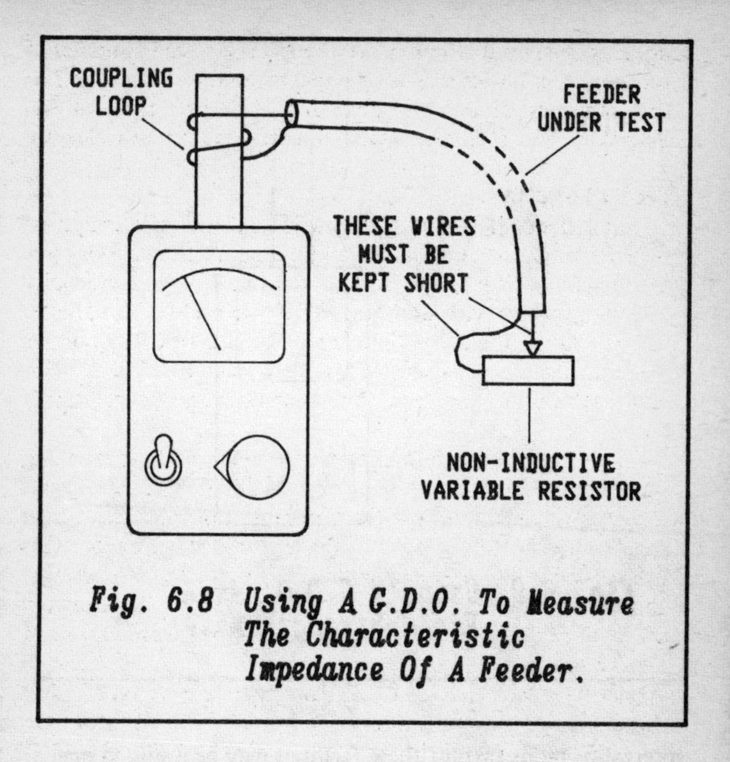


*Fig. 6.8 Using A G.D.O. To Measure The Characteristic Impedance Of A Feeder.*

be tuned to the correct frequency and the measurements and adjustments can be made to the aerial.

It should be noted that if high powers and aerial gains are involved then the aerial should not be approached when the power is on as this can present a health hazard.

**Frequency Counter**

Frequency counters are in comparatively common use in amateur radio stations these days. Their price has fallen to the extent where they are now within the budget of many people.

When looking for a counter there are a number of specifications which need to be investigated. One is the type of counter. Obviously many will measure frequency, but it is also possible to measure many other parameters such as time

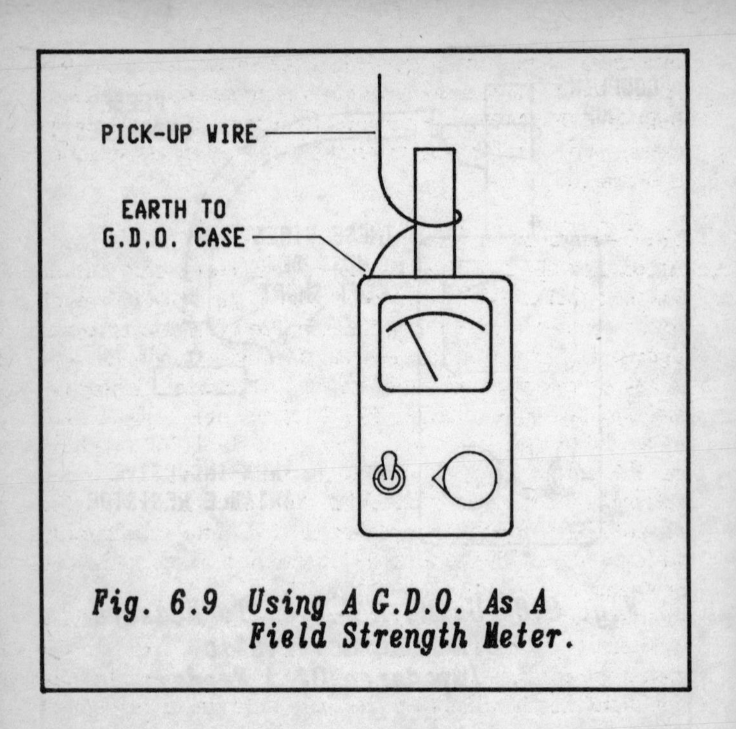


*Fig. 6.9 Using A G.D.O. As A Field Strength Meter.*

interval or ratios. Whilst these facilities may be useful in many applications a counter with these added facilities will cost more and there is no point in paying the extra if they are not needed. For most purposes within a radio shack a straight frequency counter will suffice. Another parameter which should be investigated is the top frequency limit. Normally the counter should be able to measure comfortably above the top frequency of operation which is envisaged. Even if the frequencies of operation are increased it is possible to use what is called a prescaler. In fact many prescalers are sold with counters and they divide the incoming frequency by a fixed factor (normally 10). By adopting this approach many counters can measure frequencies in excess of 1 GHz for a comparatively low cost.

When investigating the purchase of a counter it is worth bearing in mind that even though some form of frequency measuring equipment is needed in the shack, many modern receivers and transceivers have quite accurate counters already built into them. However, a good frequency counter is always useful, particularly if any transmitter or receiver construction is envisaged.

**The Oscilloscope**

An oscilloscope can be a very useful item in any radio station. The fact that it displays the waveform of the signal it is monitoring can give a whole new dimension to the measurement of signals. In fact within the electronics industry the oscilloscope is one of the most used items of test equipment. Unfortunately oscilloscopes are expensive. They will often cost £300 or more, although it is possible to pick up some quite good buys on the second-hand or surplus market. However, for the average radio amateur, whether he is interested in construction or not, it is usually possible to survive quite satisfactorily without one, and if one is needed it may be possible to use one belonging to a friend.

If an oscilloscope is required then it is necessary to have a clear idea what it will be needed to do. Extra facilities will cost a lot extra, but alternatively if the correct facilities are not there then it will not be able to perform the tasks required of it.

One of the first points to consider is the number of channels which are required. Although there are a few oscilloscopes with only one channel, most of them have two or more. A second channel is very useful as it enables two signals to be compared with one another. Possibly the input and output of a circuit might need to be displayed at the same time, or alternatively it might be necessary to look at the relative phase of two signals. Even so for most of the time only one channel will be used.

Another parameter to note is the bandwidth. Basically this reflects the maximum frequency to which the 'scope will work accurately. Some early oscilloscopes had very low bandwidths of a few megahertz. However today even the low cost 'scopes will operate up to frequencies of 50 or 100 MHz, and the more expensive ones will have much higher bandwidths. For the radio amateur a 'scope having a comparatively small bandwidth is quite satisfactory because many of the RF measurements can be made with other instruments. Even for the occasions when a high frequency measurement is required the added cost is probably not worthwhile as they are likely to be used comparatively rarely.

**Further Reading**

"Getting the Most from Your Multimeter" (BP239) by R. A. Penfold. Publisher: Bernard Babani (publishing) Ltd
ISBN 0 85934 184 4

"More Advanced Uses of Your Multimeter" (BP265) by R. A. Penfold. Publisher: Bernard Babani (publishing) Ltd
ISBN 0 85934 210 7

"How to Use Oscilloscopes and Other Test Equipment" (BP267) by R. A. Penfold. Publisher: Bernard Babani (publishing) Ltd
ISBN 0 85934 212 3

## Chapter 7

# PREPARING FOR THE LICENCE

Many people find that once they have been short wave listeners for a while they want to become licensed radio amateurs and be able to transmit. In most countries this is not too difficult, but there is usually a test or examination of some form which has to be taken. Generally this is in the form of a written examination and then if operation on the HF bands is envisaged a morse test will be required as well.

Usually these tests are not too difficult. They are planned so that most people will be able to pass. Even so a reasonable amount of time and effort will be needed to reach the required standard. In addition to this the examinations or tests can seem an insurmountable hurdle before they are taken. However, if one is determined to pass then the major hurdle has been overcome even though it may not always be easy.

### Types of Licence

Regulations about amateur licensing vary from one country to another. Different governments require different standards to be achieved before they will issue a licence. In addition to this different privileges are given. For example, the power limits which are available vary from one country to the next, as do the actual band limits to a small degree. Finally there are sometimes different grades of licence within a country which will allow access to different bands or the use of different modes.

In the United Kingdom there are four grades of licence. The first is known as the Class A licence. This allows access to all the U.K. amateur bands shown in Figure 7.1. It is only issued if the applicant has passed a theory examination called the Radio Amateur's Examination (R.A.E.) as well as a Morse Test. Then there is a Class B licence which gives access to the bands above 50 MHz and this can be obtained by passing only the R.A.E.

There are also two types of novice licence. These licences were introduced at the beginning of 1991 to encourage

*Figure 7.1 U.K. Amateur Bands*

| *Frequency Bands (MHz)* | | *Approximate Wavelength* | |
|---|---|---|---|
| 1.81 | 2.0 | 160 Metres | (Top Band) |
| 3.50 | 3.80 | 80 Metres | |
| 7.00 | 7.10 | 40 Metres | |
| 10.10 | 10.15 | 30 Metres | |
| 14.00 | 14.35 | 20 Metres | |
| 18.068 | 18.168 | 17 Metres | |
| 21.00 | 21.45 | 15 Metres | |
| 24.89 | 24.99 | 12 Metres | |
| 28.00 | 29.70 | 10 Metres | |
| 50.00 | 52.00 | 6 Metres | |
| 70.00 | 70.50 | 4 Metres | |
| 144.00 | 146.00 | 2 Metres | |
| 430.00 | 440.00 | 70 cms | |
| 1240.00 | 1325.00 | 23 cms | |
| 2310.00 | 2450.00 | 13 cms | |
| 3400.00 | 3475.00 | 9 cms | |
| 5650.00 | 5680.00 | 6 cms | |
| 5755.00 | 5765.00 | | |
| 5820.00 | 5850.00 | | |
| 10000 | 10500 | 3 cms | |
| 24000 | 24250 | | |
| 47000 | 47200 | | |
| 75500 | 76000 | | |
| 142000 | 144000 | | |
| 248000 | 250000 | | |

newcomers into the hobby. As such they offer fewer facilities as shown in Figure 7.2, but they do give an ideal opportunity to get on the air without having to reach standards quite as high as those needed for the other licences. In order to obtain a licence applicants must attend a special course and sit a simple theory examination. This will enable the novice allocations above 50 MHz to be used. Then if a fairly straightforward morse test is passed, access to all the novice allocations is given.

*Figure 7.2 U.K. Novice Licence Bands*

| *Frequency Band (MHz)* | *Types of Transmission Permitted* |
|---|---|
| 1.950 – 2.000 | Morse, Telephony, RTTY, Data |
| 3.565 – 3.585 | Morse |
| 10.130 – 10.140 | Morse |
| 21.100 – 21.149 | Morse |
| 28.100 – 28.190 | Morse, RTTY, Data |
| 28.225 – 28.300 | Morse, RTTY, Data |
| 28.300 – 28.500 | Morse, Telephony |
| 50.620 – 50.760 | Data |
| 51.250 – 51.750 | Morse, Telephony, Data |
| 433.000 – 435.000 | Morse, Telephony, Data |
| 1240.000 – 1325.000 | Morse, Telephony, RTTY, Data, Facsimile, SSTV, FSTV |
| 10000.000 – 10500.000 | Morse, Telephony, RTTY, Data, Facsimile, SSTV, FSTV |

Maximum power 3 Watts RF output or 5 Watts DC input

Further details of all the U.K. amateur licences can be obtained from The Radio Amateur Licensing Unit whose address is given in Appendix 1 along with a number of other useful addresses.

**Preparing for the Novice Licence**

As the novice licence requires the applicant to have undergone a practical course, the first step is to enrol at a suitable course. The R.S.G.B. have appointed a large number of radio amateurs willing to run these courses, and there should be a course quite close to most people. In the first instance it is necessary to make some enquiries from the R.S.G.B. who will be able to give details of the nearest courses.

The courses themselves consist of a number of worksheets all of which must be successfully completed. Each one is designed to instruct the candidate in some of the basic skills required from wiring different plugs, using multimeters and soldering to operating procedures and codes, including a bit about the morse code.

Having completed the practical course it is possible to sit the Examination. This is run by The City and Guilds and is a 90-minute multiple choice exam. The syllabus covers subjects included in the course as well as the licence conditions.

If these two sections are completed then it is possible to obtain one of the licences for use above 50 MHz. However, if any operation on the HF bands is to be considered, then a morse test has to be taken. Fortunately this is only 5 words per minute, and it is administered by the R.S.G.B. who can supply further details.

**The Radio Amateur's Examination**

If either a Class A or Class B licence is to be obtained, then the Radio Amateur's Examination has to be passed. Whilst this is somewhat more advanced than the examination for the novice licence there is no reason why most people cannot successfully pass it given some time and effort even though it may seem a large hurdle beforehand.

Like the novice licence exam, the R.A.E. is set by The City and Guilds. It may be taken either in May or December and it consists of two multiple choice papers. The first is 75 minutes long and covers the licensing conditions, transmitter interference and electromagnetic compatibility (E.M.C.). Essentially this last section deals with the interference which can arise when a transmitter is operating in the close proximity to other electronic apparatus like domestic televisions, videos, hi-fi's and the like.

The second paper only lasts for 30 minutes and covers operating procedures, simple electrical theory, solid state devices, receivers, transmitters, propagation and antennas and finally measurements.

**Learning the Theory**

When considering studying for the R.A.E. there are a number of steps which can be taken to help. The first is to find an evening course. There are a number which are run in local schools and colleges. Details of any courses can be found either from the local education authority, or alternatively the R.S.G.B. will be able to advise on their location. Courses are sometimes also run by radio clubs. A trip to the local club would soon ascertain whether they run them. Alternatively the R.S.G.B. would have details of any clubs providing this facility, and also many magazines will give details of courses in their news pages.

There are a number of good books which can help with the studies for the exam. The R.A.E. Manual mentioned at the end of this chapter is a very good book and is likely to be used as the basis of any course. In addition to this, it is also helpful to study other radio theory books as well. Some may be aimed at the R.A.E. but even if one is not it may help to extend one's overall knowledge of the subject.

With the number of books that are available for the radio amateur, it is quite possible to study for the exam without attending a course. In fact there are many successful students each year who have prepared themselves for the examination on their own. It is usually more difficult to do it this way, but still quite possible.

Apart from direct study there are a number of other ways of helping to prepare for the R.A.E. For example, constructional projects will help to give a better background to the theory. It is always better to learn by doing something than just reading it in a book. Solving some of the problems which arise during a project and then learning to use the equipment will provide some very useful experience that cannot be obtained in any other way.

Listening on the amateur bands is also very useful. Not only will it show the operating procedures which are used, but it will also highlight topics like propagation. In addition to this, it will always be possible to hear a few technical discussions which will again add to anyone's background knowledge.

**Taking the Exam**

Having studied for the R.A.E., the day for the exam itself will eventually arrive. At this late stage there is little that can be done in terms of last minute learning, but it is still necessary to prepare by being in the right frame of mind. This can take a number of forms.

The first point is to know exactly where the examination centre is, and how to get there on time. Also plan to arrive in ample time. It will certainly not help one's frame of mind if time is running short and you are not exactly sure where to go!

Once in the examination room try to keep calm. Listen carefully to the instructions given out before the examination. Then once it has started read the paper carefully. All too often marks have been given away needlessly by some point not being properly understood. If all the questions have been finished before time then read everything through again and check the answers. By doing this many people have gained valuable extra marks.

**Learning the Morse Code**

Once the R.A.E. has been taken, the next stage in obtaining a Class A licence is to take the morse test. For many people this can seem an even bigger hurdle than the R.A.E. Fortunately it is not often as big as it may seem, but it will require time and dedication to master the code sufficiently for the test.

To help in preparing for the test a number of radio clubs run morse classes. If there is a club locally then it is well worth investigating to see whether they run any classes. Even if they do not, then it is still possible to obtain a lot of help and encouragement.

In addition to set classes there are a number of other aids which are available for learning morse. For example, it is possible to buy prerecorded tapes, or ask a local amateur to help make one. Then there are items like morse "tutors" which are essentially electronic units which generate random numbers or letters. They have the advantage over tapes that the letters and numbers are generated in a random order and it is not possible to know what is coming next. When a tape is used it is surprising how few times it has to be played before one tends to learn which letters are coming next.

It can be very helpful to learn the code with a friend. In this way the impetus can be kept going. It is quite easy to put off a practice session to another day on one's own. However, with a friend there is a certain amount of loyalty which should keep the practice sessions going. Obviously it is not necessary to have every practice session together, but it is helpful to have possibly one session a week together. Another important point is to have regular practices. A quarter of an hour a day is far better than an hour and three quarters once a week.

The first stage in actually learning the code is to learn the different characters. This is best done by learning the rhythm or sound of the morse character because this is the way it will be recognised when receiving morse. There are a number of ways in which this can be done. One is to have a tape of all the characters which are to be learnt. They can be repeated a few times each before the next character is sent. Having learnt the characters it will be possible to start receiving some slow morse. Morse tutors are useful here as they enable the speed to be built up. Another source of practice is the slow morse transmissions which can be heard over the air. The R.S.G.B. and a number of other national radio societies organise a large number of these transmissions especially for people learning the code. Details appear in the magazines from time to time, but the societies will be happy to provide details. Another alternative is to start listening to some of the slower normal amateur transmissions. Eighty metres at weekends is one good source, and another good hunting ground is around the CW QRP operating frequencies.

*Figure 7.3 CW QRP Operating Frequencies*

| *Band* | *Frequency (MHz)* |
|---|---|
| 80 Metres | 3.560 |
| 40 Metres | 7.030 |
| 30 Metres | 10.106 |
| 20 Metres | 14.060 |
| 15 Metres | 21.060 |
| 10 Metres | 28.060 |

Sending morse should not be attempted straightaway. There is always a temptation to start sending as soon as possible. It should be left until it is possible to copy about eight or ten words a minute. By doing this many of the bad sending habits which would almost certainly be picked up can be avoided, and once learnt they are more difficult to correct.

The actual methods of learning morse will vary from one person to the next. However, with some perseverance it should be possible to take the morse test with a good chance of success.

## On The Air

Having passed all the tests and examinations the day will eventually arrive when the prized licence with its brand new callsign will arrive. All being well there should be some equipment available to put the new callsign on the air straightaway.

Until some experience is gained it is best not to dive into the first major pile-up heard on the bands. Whilst many people would like to have a DX station from a rare country as the first contact in the log, it is more realistic to try a few less demanding contacts first. In this way the feel of operating can be gained.

As time goes on it is worth experimenting with new bands and new modes. Only by doing this is it possible to explore all the avenues available and of interest. In particular, it is worth trying a few morse contacts. For people who have spent time in learning morse for a test, it is a little disappointing never to use it again. In fact with the pressure of the morse test out of the way many people have found that they actually enjoy morse and some actually prefer it to other modes. In addition to this it is much easier to make DX contacts using CW especially if the aerial and equipment are not large and powerful.

But whatever mode is used and whatever bands are chosen, the main object is to enjoy the hobby: its operating; its construction; discovering new aspects of radio; and helping others as well.

## Further Reading

"R.A.E. Manual" by George Benbow. Publisher: Radio Society of Great Britain
ISBN 0 900612 84 3

"How to Pass The R.A.E." by Clive Smith and George Benbow
Publisher: Radio Society of Great Britain
ISBN 0 900612 86 X

"The Secrets of Learning The Morse Code" by Mark Francis G0GBY. Publisher: Spa Publishing
ISBN 0 9512729 0

# Appendix 1

## ADDRESSES

American Radio Relay League
225 Main Street
Newington
Connecticut 06111
U.S.A.

City and Guilds of London Institute
46 Britannia Street
London WC1X 9RG
United Kingdom

Radio Amateur Licensing Unit
Post Office Counters Limited
Chetwynd House
Chesterfield
Derbyshire S49 1PF
United Kingdom
Telephone: 0246-217555

Radio Society of Great Britain
Lambda House
Cranbourne Road
Potters Bar
Hertfordshire EN6 3JE
United Kingdom
Telephone: 0707-59015

# *Notes*

# Index